Jürgen Wittmann

Zero Defects – Dead or Alive?

Introduction to Quality Management in the Semiconductor Industry

"The primary purpose of our book is to provide the "fast lane" to practical quality engineering and management in the semi-conductor industries. In line with this objective, this book is meant to be more a guide to practical quality engineering and management, rather than a scientific treatise. Although it has been written for the semiconductor technology community, it goes without saying that it is useful for almost all other industrial areas, since in a way semiconductor technology (in particular micro-electronics) is a good model case how 100% stringent QM can be implemented in practice."

Come to Berlin and study

- Mechatronics Bachelor of Engineering
- Mechatronics Master of Engineering

Learn about MEMS Technology & Materials and Quality Management

Learn for your Future!

www.beuth-university.de

Author

Prof. Jürgen Wittmann worked for fifteen years in the semiconductor and photovoltaics industry mainly in technology, materials & suppliers and quality management (production, suppliers, customers). He conducted more than seventy worldwide supplier audits. His publications focus on silicon substrate materials and quality management. Today he teaches quality management and microsystems technology at Beuth University of Applied Sciences, Berlin. He can be contacted at jwittmann@beuth-university.de and qmwittmann@gmail.com.

© Jürgen Wittmann
Beuth University of Applied Sciences, Berlin, Germany
Luxemburger Str. 10, 13353 Berlin, Germany

ISBN-10: 1986606295
ISBN-13: 978-1986606295

Zero Defects – Dead or Alive?, 1st Ed., Mar 17th, 2018

This work is subject to copyright. All rights are reserved.
Printed by CreateSpace Independent Publishing, Amazon Media EU S. à.r.l., 5 Rue Plaetis, L-2338, Luxembourg in Luxembourg

Contents

Contents III

Zero Defects – Dead or Alive? 1

1. Introduction: the term "Zero Defects" 1
2. History of Zero Defects 3
   2.1. The Beginning 3
   2.2. 50 years of Zero Defects in literature and online publications 5
   2.3. Criticism 9
3. Cost of Zero Defects 9
   3.1. Overview of Cost of Quality 10
   3.2. Models for Cost of Quality 12
      3.2.1. Traditional Model 12
      3.2.2. Alternative Models 13
      3.2.3. Proposal for a sub-dpm-Model 14
4. Industrial Product Defect Level 16
   4.1. Method of Evaluation 16
   4.2. Understanding of dpm and Related Criteria 17
   4.3. Relevance of Quality for the Supplier Evaluation 20
   4.4. dpm based Classification of Supplier Performance 21
   4.5. Other Defect related Criteria for Supplier Performance Rating 24
   4.6. Conclusion 26
5. Case Example: Automotive Semiconductor Quality 27
   5.1. Quality Level and Requirements 27
      5.1.1. Automotive Product Quality 27
      5.1.2. Future Trends 30
   5.2. Low dpm Quality Methods and Tools in the Automotive Semiconductor Industry 32
      5.2.1. General Tool: Failure Mode and Effects Analysis 33
      5.2.2. Low dpm Tools in Design 37
      5.2.3. Low dpm Tools in Development, Qualification and Ramp Up 43
      5.2.4. Low dpm Tools in Manufacturing 46
      5.2.5. Low dpm Tools in Test 48

    5.3. Learning and Consequences 50
        5.3.1. Learning 50
        5.3.2. Consequences & Proposed ZD Business Process 51
6. Summary and Conclusion 57
7. References 59

Zero Defects – Dead or Alive?

Evaluation of the Meaning and the Implementation of Zero Defects in the Industry and Conclusions

J. Wittmann, Beuth University of Applied Sciences

From the beginning "Zero Defects" has not focused on a clearly defined specific product failure rate, but was more a management tool for defect reduction. Today, in addition, quality sensitive branches predominantly consider a "defects per million" (dpm) quality level as Zero Defects, but already move on towards lower defect rates on the "defects per billion" (dpb) quality level. Conventional quality cost models still use percentage scales. So a modified cost model is proposed supporting defect levels in the dpm and sub dpm range.

Based on the evaluation of the companies' supplier quality rating schemes, however, it becomes obvious that failure rates of incoming goods in the percentage range are still accepted in the industry. Quality norms and regular quality management tools and methods are not sufficient to achieve sub dpm failure rates. In fact, additional methods and another change of attitude have to be implemented in the companies concerned.

1. Introduction: the term "Zero Defects"

Zero Defects seems to be a not clearly defined and at the same time controversial term in the Quality Management area. In 1962 Philip B. Crosby defined Zero Defects to be the performance standard by avoiding all errors and condoning no mistakes.[1] Do it right the first time!

It was considered "not just a motivational slogan, but an attitude and commitment to prevention. Zero Defect, however, does not mean that the product has to be perfect".[2] According to Crosby improvement of quality cannot be achieved by using quality tools only, but has to include organizational changes focusing on quality and has to start at the management level.[3]

---

[1] Crosby, P. B. (1962)
[2] Suárez, J. G. (1992)
[3] Zollondz, H.-D. (2011)

The Quality and Reliability Assurance Handbook "A guide to Zero Defect" defines Zero Defects as "a motivational approach to the elimination of defects attributable to human error"[1] and it is not considered a replacement for quality control.

> **Zero Defects**
> "Zero Defects is a management tool aimed at the reduction of defects through prevention. It is directed at motivating people to prevent mistakes by developing a constant, conscious desire to do their job right the first time." [2]

Later on Zero Defects used to be defined as: all parts and all characteristics which were sampled are within specification. However, this definition is not in favour of more and stable quality improvements.[3]

Zero Defects may also be defined as business practice that "can be elaborated as the commitment to complete a task within the time period decided while the agreement was done and all the required resources should be kept ready so that the goal is attained. However, the final target is to doing or executing the business practices right at the first time."[4]

From a more technical point of view, in particular when discussing product quality requirements in the highly demanding automotive industry, Zero Defects means a strategy to achieve dpm or even sub dpm quality and a quality target at the same time. The automotive industry is a high volume production industry producing millions of parts, which allows to actually prove that less than one defective part per million can be produced.

Hence, Zero Defect is now frequently associated with more concrete figures regarding failure, e.g. at Altera (now part of the INTEL Corporation):

> "As a cornerstone of that quality effort, we pursue a zero-defect strategy that has successfully driven returndefect-rates down to < 1 defect per million" [5]

---

[1] Fouch, G. E. (1965)
[2] Halpin, J. F. (1966)
[3] Sullivan, L. (1984)
[4] Ardianto, M., (2014)
[5] Mazotti, W., (2016)

In the next step, the automotive customers, car makers and their supply chain, expect to receive literally Zero Defect products from their respective suppliers, i.e. customers' expectations increased **from dpm down to single events**.[1]

Apart from the automotive industry, there are, of course, many other branches which require a very high level of product quality. Those products frequently must have a very low fail rate at very long operating times. They might be operated at harsh environmental or use conditions or in safety critical applications and may be difficult to be replaced, e.g.

- Space and Aviation
- Medical Equipment and intrabody devices
- Energy systems
- Underwater cables and communication infrastructure in remote areas
- Defense
- ...

In the case of a lower production volume, however, it is obviously much more difficult to provide evidence for defect free products in the sub dpm range. Thus, new and different approaches and methods are needed to achieve and to prove low defect production. Examples for "guaranteed" reliability approaches in highly demanding medical applications are provided by Dyconex.[2,3]

2. History of Zero Defects

2.1. The Beginning

The development of Zero Defects is mainly credited to Philip B. Crosby, a quality control manager at the Martin Company, who was in particular involved in the Pershing missile program. From the beginning the program had a tight schedule (23 months to first launch) and clear goals such as to prevent defects in production, no adjustment of hardware afterwards, no further development in the production phase and the establishment of a new standard regarding quality performance in the industry.[4]

---

[1] Naransimhan, R. (2017)
[2] Klein, H.-P. et al. (2015)
[3] McNulty, J. C. et al. (2016)
[4] Crosby, P. B. (1964)

The sheer size of the project, however, implies, that there were many others, e.g. G. T. Willey, E. J. Cottrell and J. F. Halpin, involved in starting and executing Zero Defects.[1]. Several activities were started from the beginning to achieve the goals, e.g.:[2]

- Development and use of workmanship specifications
- Establishment of training for workers in order to learn required techniques
- Introduction of a design review prior to release of production
- Close supplier involvement including supplier training
- Introduction of detailed inspection and test plans
- Development and use of quality procedures
- …

Despite all the activities during the program it turned out that all the missiles delivered to the customer suffered from minor or major, partially even critical, defects. More motivation of the people did not lead to further quality improvement. After some time it became obvious that the root cause for not reaching defect free missiles was that the workers did not identify themselves with the company's quality targets. The task of the management was then to promote the worker's desire to perform his task without mistakes, right the first time by motivation and awareness for the importance of the respective task by giving them the attention required.

The management had to show to the company that Zero Defects was actually their own target without any excuses.

> "So if you are satisfied with 9 defects per missile, or tractor, that's what you'll get." [3]

As a consequence early on Zero Defects was associated with the following activities and approaches like:[4]

- Management Commitment
  The company management has to be committed to Zero Defects and to provide support and directions in order to acknowledge the contributions of the employee

---

[1] Halpin, J. (1966)
[2] Crosby, P. B. (1964)
[3] Crosby, P. B. (1964)
[4] Fouch, G. E. (1965)

> Planning

A Zero Defects program requires appropriate planning. This includes for instance
- The formulation of objectives
- Identification of the main targets and
- Setting numerical goals

Other items, e.g. procedures and reporting structures, need to be planned ahead of time, too.

> Organization

A Zero Defects program needs to have an administrator being responsible for the program and being part of the company's staff at an appropriate level of the hierarchy.

> Error-Cause-Removal

Very soon it became obvious, that many defects are not caused by human error, but by environmental conditions. Those have to be addressed separately.

These activities were not absolutely new, but the combination of all of them and the consequent compliance with the given rules and procedures made the Zero Defects programs successful.

Nowadays leadership and commitment, planning, organizational roles, responsibilities and authorities as well as the elimination of the causes of nonconformities are important parts of the relevant quality management standards, e.g. ISO9001:2015.

Conclusion:

The above mentioned activities, which used to be part of Zero Defects programs in a comparably small number of companies, are today common activities in every ISO9001:2015 certified company all over the world.

2.2. 50 years of Zero Defects in literature and online publications

During the first years Zero Defects programs were very popular, primarily in the defense and aerospace industry. Substantial improvements were reported, e.g. by Martin (54% defect reduction in two years), General Electric (2 million US$ savings in rework and scrap cost within two years) and Sperry Corporation (54% defect reduction in one year), and thousands of company planned or implemented Zero Defects programs at that time.[1]

However, the information about whether or not Zero Defects is still appreciated in the industry today is very controversial.

On the one hand, and this is in line with the final sentence of the previous section, quality methods, tools and systems like Total Quality Management (TQM), Malcolm Baldrige or other quality awards, statistical process control (SPC), ISO9001 and Six Sigma, incorporate or replace the ideas of Zero Defects: "now the zero-defects approach is practiced by only a few companies. Except for those who profit from it, zero defects is forgotten".[2]

On the other hand, Zero Defects quality, which is driven by customer requirement or by safety requirement, e.g. in the automotive industry and in the aviation industry, "is implemented all over the world".[3]

This discrepancy was the reason to investigate the number of publications over the years containing the term "Zero Defects" in combination with other quality relevant terms as a measure of relevance for the industry.

The literature search is based on two different sources, the freely accessible web search engine Google Scholar and the Bielefeld Academic Search Engine (BASE) for scientific documents (https://www.base-search.net/). The focus of this search is the identification of changes over time. The fact that Google Scholar provides a number of hits per term which is at least an order of magnitude higher than with BASE is not subject of this investigation.

The following combinations of terms were used for the literature search for Zero Defect:

ZDP BASE : "Zero Defects Program" for BASE
ZDP G'Scholar : "Zero Defects Program" for Google Scholar
ZDQM BASE : "Zero Defects" AND "Quality Management" for BASE
ZDQM G'Scholar : "Zero Defects" AND "Quality Management" for Google Scholar

Publications with the term "Quality Management" (QM BASE, QM G'Scholar) and the total number of publications in the BASE data base were used as a reference.

First of all fig. 1 shows a significant growth of the number of publications in BASE of a factor of 26 over the last 55 years. Most likely the number of publishers has increased over time. Whether or not older publications were not considered to be inserted in the database is not known. The search in Google Scholar over periods of five years

---

[1] Halpin, J. F. (1966)
[2] Crosby, D. C. (2006)
[3] Wang, K.-S. (2013)

unfortunately yielded no reasonable number of publications and was therefore not included here. Compared to the total number from BASE the number of publications including the term "Quality Management" in both literature searches follows a much steeper increase which can be interpreted in a way that the importance and/or use of quality management increased significantly within this time frame.

The term "Zero Defects program", however, is almost invisible in BASE and, with a peak from 1966 to 1970 and a subsequent decline from 1970 to 1985, stabilizes below one hundred publications per five year period in Google Scholar. In relation to the number of overall quality management related publications the share of publications with "Zero Defect program" decreases over time. The reason for this is assumed to be the introduction of other quality tools, methods and programs as explained before. A separate Zero Defects program seemed not considered to be necessary for the companies, because the methods of ZD are applied in daily life and new kinds of quality improvement programs were put into focus.

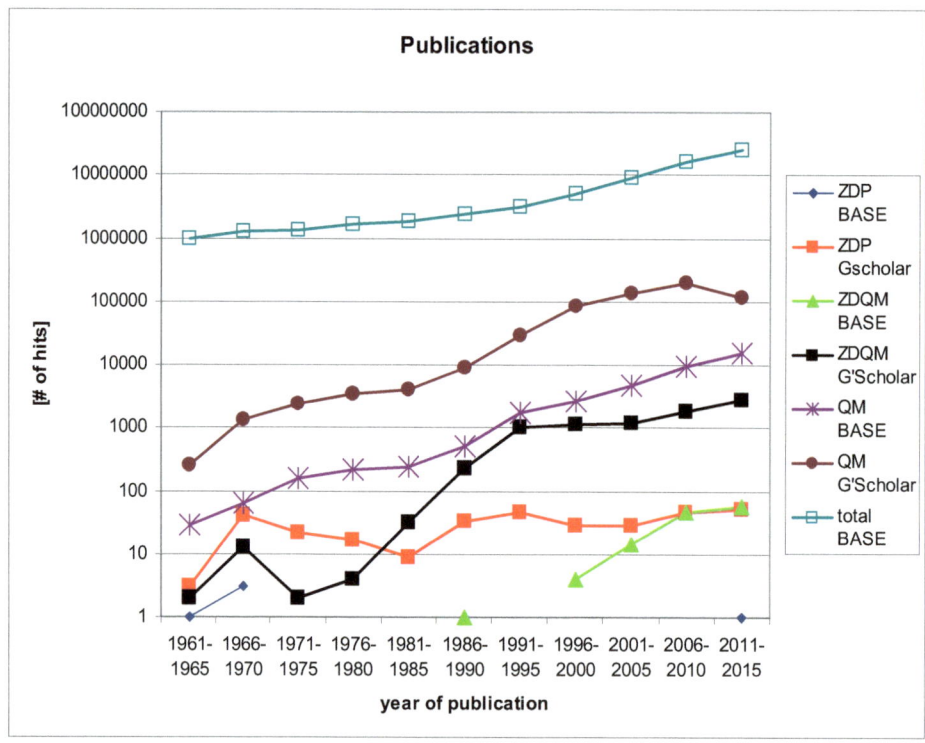

Fig. 1 Number of hits for different Zero Defect related terms in Google Scholar and BASE

Later in time, starting in 1980 in Google Scholar and in 1995 in BASE, the number of publications containing "Zero Defects" in combination with "Quality Management" increases significantly which shows an increasing importance of Zero Defects. This is in accordance with the increasing importance of actually defect free products in the automotive, aviation and other industries. Hence Zero Defects experiences a "come back", but the meaning has shifted to some extent.

In order to identify industrial branches where Zero Defects appears to be topical today, another search on Google was performed using the search term "Zero Defects" only. From the first 500 hits all hits related to consulting, academics, non-commercial organizations etc. were removed and a focus was put on commercial companies with relevance for the respective industry.

Being aware, that this is a more pragmatic approach with weaknesses in the identification and assignment of specific branches, nevertheless it shows, as expected, that with 35% the biggest portion of hits was found at companies in or delivering to the automotive industry (fig. 2). This includes several automotive semiconductor companies. "Aerospace & defense" and "medical, health & life sciences, in descending order, achieve in total 28% and represent the production of safety critical products with very high quality requirements.

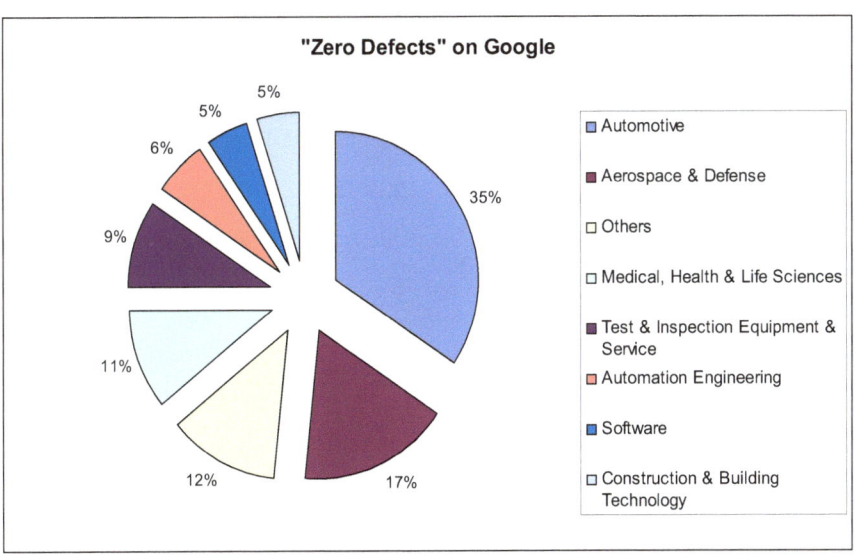

Fig. 2 Industrial Sectors with hits on the term "Zero Defects"

"Others" comprehend several branches and services with one hit only. However, also supporting branches like "Test & inspection equipment and services", "automation engineering" and "software" obviously adopted the term "Zero Defects" to express their understanding of what quality means to them or at least as an advertising point.

2.3. Criticism

From the beginning, however, Zero Defects drew a lot of criticism, also from pioneers of Quality Management like W. Edwards Deming. Both the potential negative impact on employee's motivation and the presumably increasing cost to prevent any single defect are, at least partially, assumed to be the reason for the decline of Zero Defect in the mid 1970s.

Zero Defects programs aim for changing the company's attitude towards avoiding any kind of mistake. On the shop floor, however, the term "Zero Defect" implies that no mistakes are allowed anymore and making mistakes might have consequences in the future. Making mistakes is perceived as something very human and thus to be unavoidable. Being aware of this unachievable goal of no defects whatsoever, this may lead to frustration and a reduced productivity among the workforce of the company. What has presumably been meant by the originators of the Zero Defect concept is that technical solutions have to be put in place to render mistakes impossible, such as replacing manual reading of numbers and typing them in on a key board should be replaced by a bar code reader or similar failsafe transmission of data.

> „**Quality Guru Deming** believed that slogans like "Zero Defects" are actually counterproductive and may deemphasize the culture and tools associated with continuous improvement. "[1]

The cost issue associated with Zero Defects needs a somewhat deeper discussion which is therefore addressed in the next session.

3. Cost of Zero Defects

When looking at cost for Zero Defects it is important to understand, that from the beginning of industrialization the way to ensure product quality used to be massive quality control. The American defense industry, for example, which always requires

---
[1] CSQA (2017)

reliable and defect free products, employed "250,000 people controlling processes and checking products before delivery"[1] at the beginning of the Pershing program. In addition more than 50,000 government quality inspectors were needed to ensure quality. This entire workforce obviously drove quality cost significantly. At the same time the actual product quality target was to reduce "its rate of defects 50 percent below the acceptable quality level (AQL)"[2]. Talking about defect rates on AQL level, i.e. low percentage down to sub-percentage, however, is far away from defect free products. From today's point of view, obviously the introduction of quality management methods resulted in a reduction of the quality inspection workforce, hence in a reduction of cost of quality.

If Zero Defects is interpreted quantitatively as defect free product or process not only in the range of "sub-dpm down to single events", but without any allowed defect then common sense implies that "no defect at all" leads to, at least, very high cost for avoiding defective parts from reaching the customer. In order to make this more clear cost of quality is explained in the following sections.

3.1. Overview of Cost of Quality

"Quality costs are a measure of the costs specifically associated with the achievement or non-achievement of product or service quality."[3] For other definitions of quality cost the reader is asked to refer to the literature.[4]

Typically cost of quality is classified into the following categories:

> ➢ Prevention Costs
> Prevention costs are costs for all activities which are required to avoid poor quality in products, e.g. defects. Those activities include for instance
> - Product and process quality planning
> - Supplier management, e.g. selection, audits, assessments, etc.
> - Training
> - Quality improvement teams and meetings
> - etc.
>
> ➢ Appraisal Cost

---

[1] Halpin, J. (1966), p. 10
[2] Halpin, J. (1966), p. 12
[3] Pyzdek, T., Keller, P., (2012)
[4] Kim, S, Nakhai, B., (2008)

Appraisal costs include costs for the actual measuring and evaluating of products with the goal to ensure product conformance to standards and requirements. This includes activities like
- Incoming inspection
- Process inspection
- Final inspection (or test)
- Audit activities

> Internal Failure Costs

Handling and disposition activities of non-conforming products lead to additional cost in production, namely internal failure costs. Failure costs are internal when they arise prior to delivery. They may me caused by the following activities:
- Scrap of non-conforming products
- Rework and re-inspection of non-conforming products
- Downtime of equipment
- Material quality review activities
- Downgrading of material
- Lost production capacity in bottleneck equipment due to the disruption of the normal material flow
- Loss of reputation and customers
- etc.

> External Failure Costs

Of course, product failures may also arise after product delivery to the customer. In this case we talk about external failure cost. Various activities may be driven after an external failure occurred, e.g.
- Handling of the customer complaint, including analysis activities
- Rework and potentially retest of returned parts
- Warranty claims & charges

In this context "defects" are non-conformances with specified product requirements or the non-fulfilment of customer expectations. In order to enable a zero-defect product, i.e. no defective part reaches the customer, activities for prevention and appraisal have to be initiated which drive additional cost of quality. Hence for the discussion of Zero Defect cost we have to distinguish between total quality cost, i.e. costs for prevention, appraisal <u>and</u> failures, and costs to avoid defective parts at the customer, i.e. prevention and appraisal costs. In the end, when it comes down to profitability, company's decisions will be based on total quality cost.

3.2. Models for Cost of Quality

3.2.1. Traditional Model

The traditional model for cost of quality consists of three contributors: appraisal, prevention and failure cost. The underlying assumption in this model is that there is a cost optimum at a minimum remaining failure rate, i.e. defect free products are not achievable from a cost point of view (fig. 3).

Improvement of product quality by means of additional appraisal and prevention activities would inevitably lead to high cost.[1] This approach is often subject of criticism. For instance, it needs to be clarified if optimum is defined as an economic optimum or a quality level optimum. Both could be justified.[2] Also quality induced increase of the company's earning or decrease due to poor quality and reputation is neglected in the traditional model.[3]

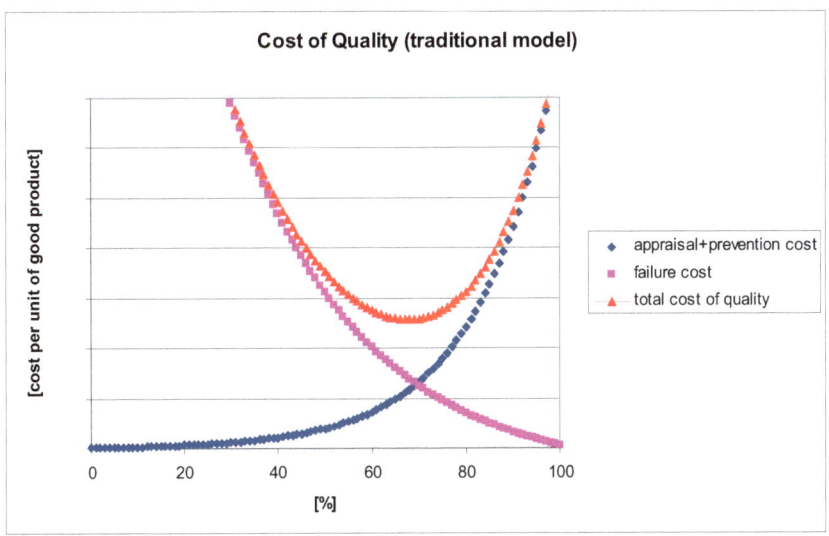

Fig. 3 Traditional Model for Cost of Quality

---

[1] Jochem, R., Raßfeld, C. (2014)
[2] Schiffauerova, A., Thomson, V. (2006)
[3] Jochem, R., Raßfeld, C. (2014)

The traditional model is not in agreement with observations in the industry.[1] New technologies enable better quality products at lower cost. Early cost based criticism of Zero Defects is assumed to be mainly caused by the traditional view on quality cost.

3.2.2. Alternative Models

Modern Model

Taking into account experience from the industry, a new – modern – cost of quality model came up with a limited increase of the appraisal & prevention cost towards a defect free product. At the same time, with a decreasing number of defects, the failure cost comes down, which results in a cost optimum right at 0% defects. This model obviously allows Zero Defects production with reasonable cost (fig. 4).

Dynamic Model

Freiesleben proposed a dynamic model which considers technological progress and learning from previous defects and non-conformances. For a given quality level, e.g. in different companies, in different branches, or at different points in time, an optimum point regarding quality cost is achievable (fig. 5).

For more information about Cost of Quality models please refer to the literature.[2] The dynamic model obviously has the potential to support defect free production at acceptable cost.

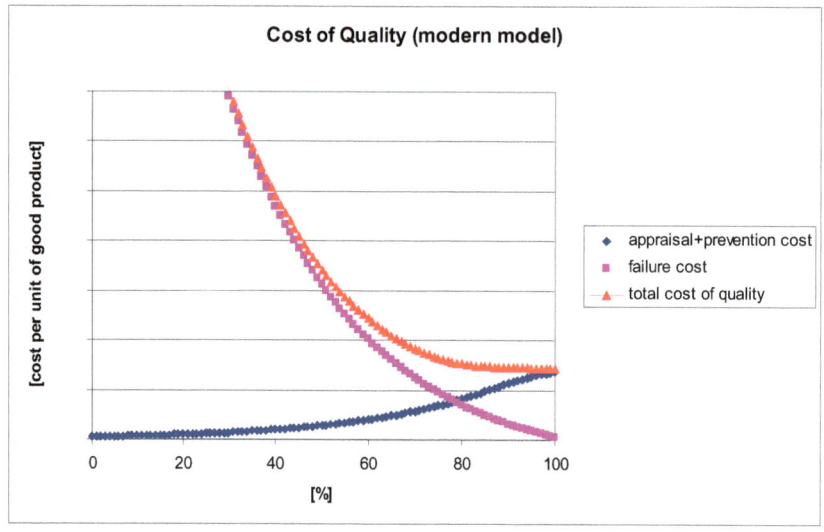

Fig. 4 Modern Model of Cost of Quality[3]

---

[1] Vaxevanidis, N. M., Petropoulos, G. (2008)
[2] Trehan, R. et al. (2015)
[3] Trehan, R. et al. (2015)

3.2.3. Proposal for a sub-dpm-Model

Trehan concludes, that "it is difficult to find a generic model that fits different kinds of industry. Further, it requires expertise to apply the given model to an industry".[1] Hence in this sub section a proposal shall be made for a generic Cost of Quality model which can be applied to technology driven companies with product quality on the dpm and sub-dpm level, e.g. the automotive semiconductor industry.

In this context the above described Cost of Quality models are somewhat restricted in two ways:

(1) Due to the limited number of products produced evidence for reaching a 100% defect free quality performance is not possible. In other words, statements about cost of quality at Zero Defect are of a very theoretical nature.

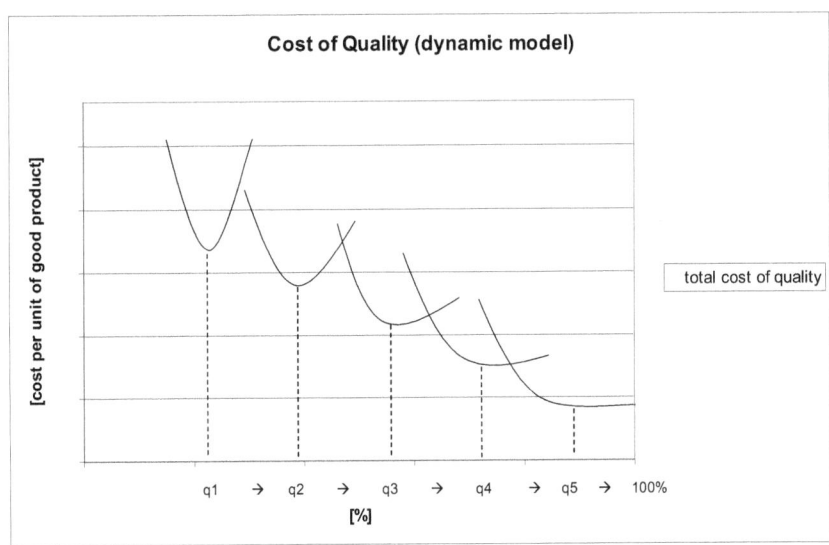

Fig. 5 Dynamic Model

(2) The description of the quality level using a percentage figure is not adequate to a dpm and sub-dpm quality level. Over the years the industry's quality level moved from hundreds of dpm down to single digit dpm und keeps moving beyond that. Hence product quality improved by a factor of at least 1000 in that time frame and all of this happened within the final 99 to 100 percentage!

---

[1] Trehan, R. (2015)

The conclusions for the dpm cost of quality model out these arguments are to switch to a logarithmic scale for the dpm rate (fig. 6). In this figure 10^6 dpm equals 0% perfection on the left hand side, while on the right hand side of the x-scale there is no actual limit. It represents a moving target and shifts towards smaller figures as the technical capabilities improve.

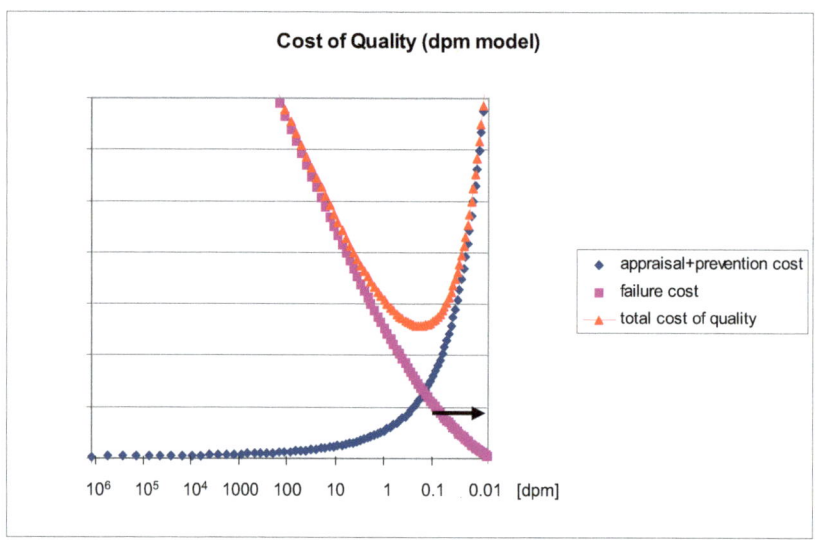

Fig. 6 Generic Cost of Quality model for dpm quality levels (example for "old model approach"). Please be aware of the log[dpm] scale as compared to the usual [%] scale

This leads to several conclusions:

> There is no provable limit like "Zero Defect", i.e. 100% conformance. Zero Defect is herewith a "moving target" or an asymptotic target, e.g. from dpm down to dpb level, and evidence for it depends on a sufficient volume of production. But then a shift of the point of view becomes necessary:

> **Think in provable dpm or even in single events rather than in un-provable perfection**

> Depending on the technical level the company has achieved there is still an optimum of the quality cost. Going beyond the regular technical capability

drives additional cost. The rise of the appraisal and prevention cost beyond the quality cost optimum, however, is not clear yet and needs further investigation.

> The dpm model can be "calibrated" using real data for cost as well as for the defect level based on the company's quality and cost data.

Therefore, by defining Zero Defect to be the lowest achievable and verifiable defect rate, the Zero Defects concept makes technological and economic sense.

4. Industrial Product Defect Level

Many industrial customers, i.e. customers receiving materials, semi-manufactured products and parts for further processing, state in their publications and quality policies that they target for Zero Defects for their products.
Consequently, they frequently expect their suppliers to deliver defect free material and parts and to aim for Zero Defects as well. From own long-time experience in the semiconductor industry however, customers are often aware that the production of defect free products is very difficult and might increase cost and tend to accept an agreed upon level of defects.

4.1. Method of Evaluation

Unfortunately for third parties it is not possible to access the actual defined and mutually agreed dpm levels within those agreements. However, in some cases and in a more indirect way, the accepted level of defects can be found in the company's supplier quality manuals or other supplier related documents.
Many supplier manuals and supplier quality manuals etc. contain a chapter about the evaluation and monitoring of the supplier's quality performance. This normally includes criteria related to logistics performance, cost performance and, of course, quality performance, and classifies the supplier by means of a point based system into different levels like "preferred", "acceptable", "on probation" and "not acceptable".

Via the internet the companies' homepages were scanned for supplier related documents and within those documents the way of the company's supplier performance rating system was examined. If the performance rating system contained a requirement expressed in dpm or at least fractions of percent and if this dpm requirement was quantified, then the company was listed for further investigation.

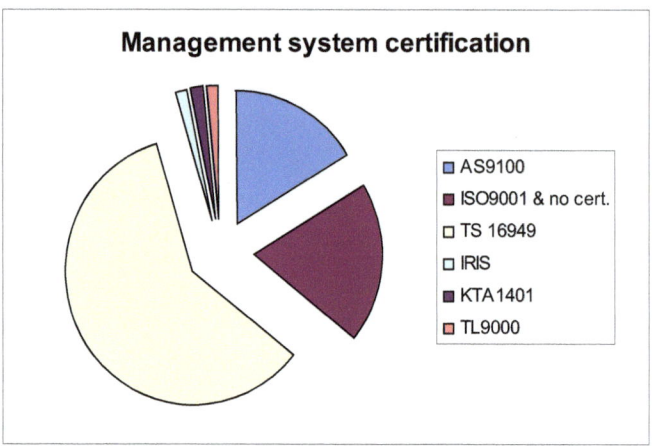

Fig. 7 The majority of documents with clearly defined dpm targets for their suppliers originate from TS 16949 certified companies

Eventually, 59 figures regarding classified dpm levels from different companies could be collected. Many companies participate in different branches like electronics, automotive, industrial and so forth, which require different levels of product quality and in some cases also different kinds of quality management system certification, e.g. ISO9001, TS 16949 and AS9100. Two companies which each having an automotive and an aerospace & defense business unit are certified according TS 16949 and AS9100 at the same time, only one company did not have any certificate. More than half of the documents with clear dpm targets for the supplier rating originate from the automotive industry (fig. 7).

4.2. Understanding of dpm and Related Criteria

In addition to the logistics performance, cost performance or others the supplier's quality performance is generally one contributor for the evaluation of the overall supplier performance.

Depending on the company the quality performance consists of the number of defective parts as compared to the number of received parts or does include other quality related indicators. In the case of MAN Nutzfahrzeuge Gruppe (MAN utility vehicle group), for instance, the supplier's quality performance is calculated using the following weighted contributors[1]:

> - Failure rate, field fails   30 %
> - Restitution 0-km   10 %

---
[1] MAN Nutzfahrzeuge Gruppe (2006)

- Serial quality   30 %
- First article quality 10 %
- Audit result   10 %
- Missed deadlines   5 %
- Performance   5 %

It is assumed that those indicators are defined in detail in the respective internal procedure. Unfortunately, this is not the case in the available presentation.

Alex Products Inc., for example, an automotive supplier, includes the following contributors for the calculation of the overall quality performance:

- Quality PPM
- PPM Improvement
- On-Time Delivery
- On-Time 8-D
- On-Time PPAP
- Repeat Issues
- Customer Issues

The respective internal procedure[1] defines in detail how many points to deduct from the top performance level for what kind of deviation.

Hence it is obvious that the dpm performance is only one perspective to judge a company's quality performance.

In addition, the definition of the meaning of "defects per million" is often not so clear from the procedures. A typical definition is

> "Total Number of Defective Parts Found / Parts Delivered in the Month x 1,000,000"[2]

or

> "Total Part Quantity Rejected X 1,000,000 = PPM Figure"[3]

---

[1] Alex Products, Inc. (2015)
[2] Sage Automotive Interiors (2013)
[3] Norgren (2015)

However, this kind of definition does not explain where in the process flow the defect was detected, e.g. at incoming inspection, during production or even at the final customer and what is included in this figure.

Other companies clearly define the origin of the fails as the basis for their calculation. For example, NCR calculates the supplier's dpm level "based on the number of failures of Supplier parts observed at NCR manufacturing plants"[1], Kärcher and Littelfuse use the number of fails from incoming inspection up to the customer:

Kärcher:

$$DPPM = (F_{Incoming} + F_{Assembly}) \times 1{,}000{,}000 \,/\, \text{quantity delivered}[2]$$

with $F_{Incoming}$ = number of objections during incoming control
$F_{Assembly}$ = number of objections during assembly and other processes up to product use

Littelfuse:

$$DPPM = (\text{\# of defective parts found in incoming control and process and Customer}) \,/\, \text{\# of total parts received}[3]$$

Whether or not fails from the customer and/or from the production or assembly site are included in the overall number of fails of course does have an influence on the dpm rate. However, from own experience differences in dpm rates due to the underlying database can be assumed to be within the same order of magnitude.

In particular if the dpm rate includes fails at the customer a clear definition for "customer dpm rate" is required. 0 hour or 0 mileage fails should be treated in a different way than fails occurring after several years under use conditions.

In addition to the uncertainty regarding the database also the time period varies from company to company. Periodic supplier performance measurements are conducted on a quarterly, semi-annual or annual basis.

---

[1] NCR (2015)
[2] Kärcher (2010)
[3] Littelfuse (2014)

Hence for the following section it should be kept in mind that the definition of the dpm figures is not absolutely consistent in between the companies and a more extensive investigation taking these circumstances into account needs to be conducted.

4.3. Relevance of Quality for the Supplier Evaluation

In order to estimate how important quality is for the industrial customers in this study, the relevance of quality in their respective supplier performance evaluation system was determined and assigned to different groups of certification. All original documents are listed in the literature list.

Being aware that other items like timely delivery of goods and cost of products etc., are also very important from a customer's point of view, it is obvious that TS 16949 certified companies tend to put more focus on quality than others. 25% of the TS 16949 certified companies weigh quality with more than 50% in their supplier performance rating procedure (fig. 8).

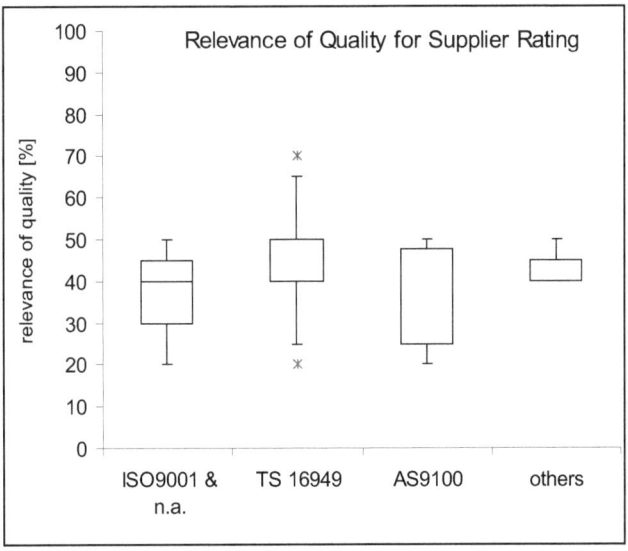

Fig. 8 Weight of quality in the supplier performance rating procedure

However, also companies with non-automotive certifications show fairly high values, which shows that companies which are aware of ppm failure levels do have a strong awareness for the importance of quality. Whether or not this is true for the majority of ISO9001 certified companies has not been subject of this evaluation.

4.4. dpm based Classification of Supplier Performance

For the judgement of the supplier's dpm performance the respective company procedures contain a table showing the number of points or percentages to be deducted from the maximum score depending on the dpm rate. In most cases a dpm range from x (minimum value) to y (maximum value) determines how many points to deduct.

In this evaluation the lowest dpm figures correspond to "class 1", the second best figures correspond to "class 2" and so forth. The "worst" class is defined by the dpm rate leading to the maximum deduction of points. Only a few companies have more than 9 classes which is why this evaluation was limited to 9 classes.

Fig. 9 shows the maximum dpm levels for the respective classes. It shows clearly that many companies use the term dpm in their supplier performance rating procedure, but have not left the % - range, i.e. end up above 10,000 dpm (1%) fail rate. For instance, two companies on the left side of fig. 9 end up with 40,000 and 35,000 dpm in the worst class, i.e. these suppliers with a product failure rate of 4.0 % and 3.5 % still receive points in their supplier performance rating. Some of those companies ask for zero defects in their supplier manual or claim to have this target on their homepage, but obviously accept failures in the percentage range.

> **The understanding of what is an acceptable level of defects seems to be very different from company to company**

"Real 'Zero Defect'" companies are assumed to accept only up to 1,000 dpm, i.e. 0.1 %, in the worst class. Fig. 10 shows the respective classification and, as in fig. 9, two groups of rating scheme become obvious: one group requires fairly low dpm rates for a class 1 classification, but allow high dpm rates above 200 dpm in class 2 and 3 already and end up at hundreds of dpm. The other group starts as low dpm rates, too, but the observed slope towards higher dpm rates in subsequent classes is much smaller. Several companies allow a maximum dpm rate of 200 dpm or below.

When looking at the highest accepted dpm rate, i.e. the last class with points adding to the supplier performance, the automotive industry by far demands the lowest dpm rates from their suppliers (fig. 11).

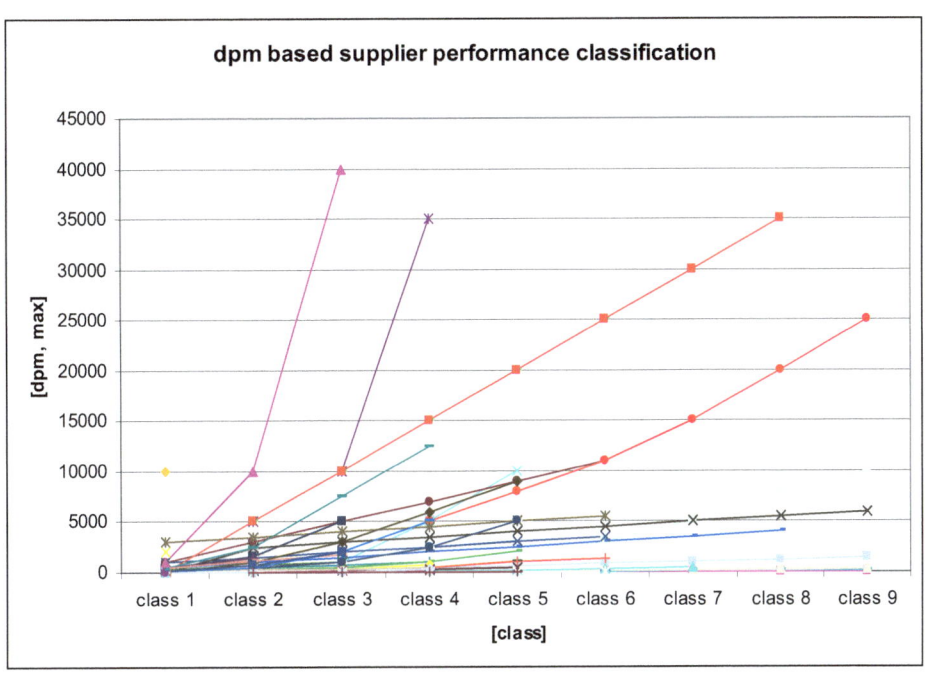

Fig. 9 dpm based supplier performance classification for all evaluated companies

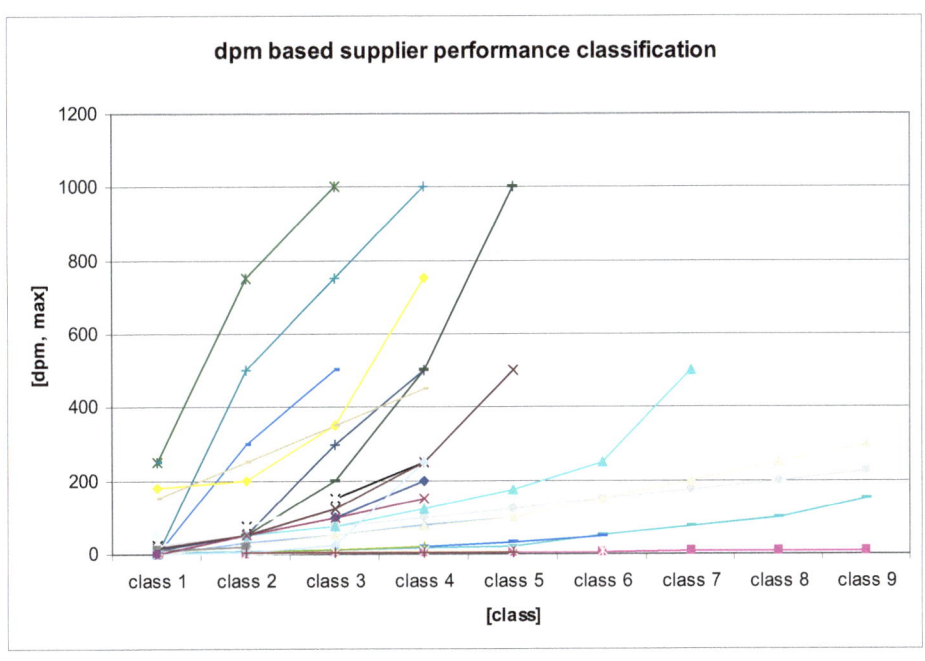

Fig. 10 dpm based supplier performance classification for "low dpm" companies

Several companies claim to have the dpm rate as one criterion for the supplier performance rating, but do not include concrete figures in their procedures. Typically they refer to a specific customer-supplier agreement for different products[1,2].

These statements reveal a more sophisticated view on the supplier's dpm performance, but obviously those could not be included in this study. However, a few companies documented different acceptance levels for different product groups in their procedures. All of them are TS 16949 certified. Fig. 12 shows class 1 to class 4 accepted dpm rates for "high quality", typically automotive parts suppliers, and "low quality", typically raw material suppliers.

In the "high quality" group the median values in classes 1 to 4 are 0, 75, 150 and 190 dpm.

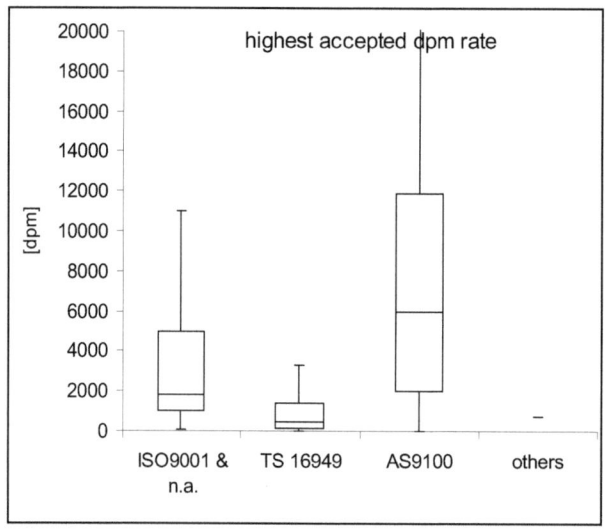

Fig. 11 Maximum dpm rate accepted by industrial customers

---

[1] Voith (2012)
[2] Kärcher (2010)

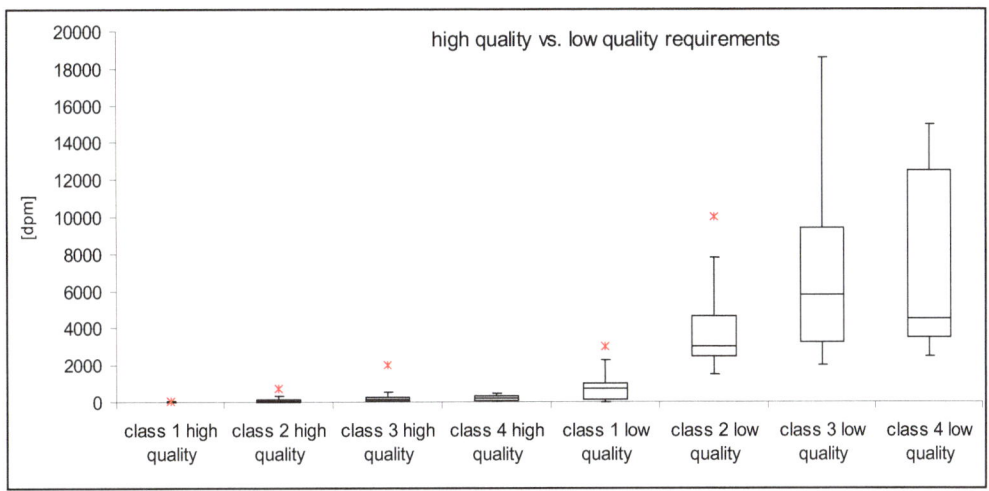

Fig. 12 Different requirements regarding dpm performance depending on the product group supplied

4.5. Other Defect related Criteria for Supplier Performance Rating

In addition to sheer dpm figures other defect related criteria are often used as a basis for the supplier performance rating, e.g.

- the severity of the fail & complaints
- reoccurrence of fails
- incidents & value

Severity of fails & Complaints

Several companies include the severity of fails, i.e. the impact on production or on product quality, in their supplier rating method. Australian Arrow, for instance, added an additional category for point deductions due to the severity of fails[1]:

- minor defect: 1 point deduction
- major defect: 2 points deduction
- critical defect: 10 points deduction

---

[1] Australian Arrow (2013)

Examples are given for the classification. Major defects are defects leading to corrective action requests or requests for 8D reports, critical defects are defects resulting in line stops or similar.

In addition to the dpm performance many companies subtract points from the maximum score for the number of complaints. BorgWarner, for example, subtracts the following number of points from the maximum 20 points for complaints within a 6 month period[1]:

No. of complaints	Deduction [no of points]
0	0
1	2
2	4
3	6
4	10
5	15
≥ 6	20

Table 1 No of complaints leading to a deduction in the supplier rating

One complaint could comprise several defective parts. A similar approach can be found e.g. at Schaeffler[2], AGM Automotive[3], Littelfuse[4] etc.

Reccurrence
Assuming an efficient quality management system on the supplier's side a defect with a particular root cause must not happen again if the respective 8D project resulted in a removal of the root cause or in an adequate additional test etc.
This is the reason for some companies to treat recurrences particularly stringent by a significant reduction of the score.[5,6]

Incidents & Value
Several companies like AGM, Littelfuse, Shilo, Volvo etc. include incidents, value or both in their supplier performance rating system.

---

[1] BorgWarner (2015)
[2] Schaeffler (2006)
[3] AGM Automotive (2015)
[4] Littelfuse (2014)
[5] BorgWarner (2015)
[6] Kirchhoff Van Rob (2016)

Volvo, for instance, combines four defect related categories to one quality performance indicator "QPM" (quality performance measurement), because "The QPM has proven to provide
a better indicator of supplier performance than by using PPM alone"[1].
The Volvo supplier rating system consists of

- defects per million
- number of inspection reports raised
- number of non-conforming parts
- volume value (cost of returned material / value of entire delivery)

In particular the number of non-conforming parts, which is independent of the number of purchased parts, reaches into a sub dpm requirement at high purchase volumes.

4.6. Conclusion

In many companies the acceptable product defect level is measured in "defects per million". However, the way the evaluation was performed does not reveal the overall share of companies using this term. In some more detail the following statements can be made:

- The majority of documents with clearly defined dpm targets for their suppliers originate from TS 16949 certified companies

- The definition of dpm figures is not consistent in between the companies, e.g. regarding
  - the sources of the data
  - different acceptance levels for different product groups (all of them TS16949 certified)

Despite the fact that the companies chosen for this evaluation often claim to aim for zero defect also on their supplier side, the spread of accepted dpm levels spans over several orders of magnitude.

- several companies using the term dpm in their supplier performance rating procedure practically still think in percent (up to 40,000 dpm being accepted)

---

[1] Volvo (2016)

> many companies ask for 0 or one digit dpm rates from their suppliers to be class 1, but accept thousands of dpm in class 2 or 3 and higher

> only a small number of companies takes the defectiveness serious enough to allow a maximum of 200 dpm for the "worst" class supplier rating. All of them are TS 16949 certified companies.

Many companies realize that the exclusive rating of the dpm performance is not sufficient for a holistic view on the supplier's product quality performance. Therefore they add other defect related categories to their supplier performance rating procedure like

> the severity of the fail & complaints
> reoccurrence of fails, in particular after executing the respective 8D procedure
> single incidents
> value of the affected material as compared to the entire material value

Hence it is obvious that the dpm performance is only one perspective to judge a company's quality performance.

5. Case Example: Automotive Semiconductor Quality

Due to the apparently very high importance of product quality, i.e. low or zero defects, to the automotive industry, the following sections focus on automotive – and in particular – on automotive electronics quality.

5.1. Quality Level and Requirements

5.1.1. Automotive Product Quality

Apart from the 1960s, in Germany the mileage per car and year (passenger cars) is more or less stable at 12.000 to 14.000 km / year (fig. 13). Reasons for ups and downs and more details about the share of "Diesel" cars, overall economic situation, number of cars per household, age of cars etc., which do have an impact on the yearly mileage per car can be extracted from the literature indicated. For instance, the way to calculate the mileage figures changed from 2005 to 2006[1], which explains the observed step in fig. 13.

---

[1] Kunert, U. et al. (2012)

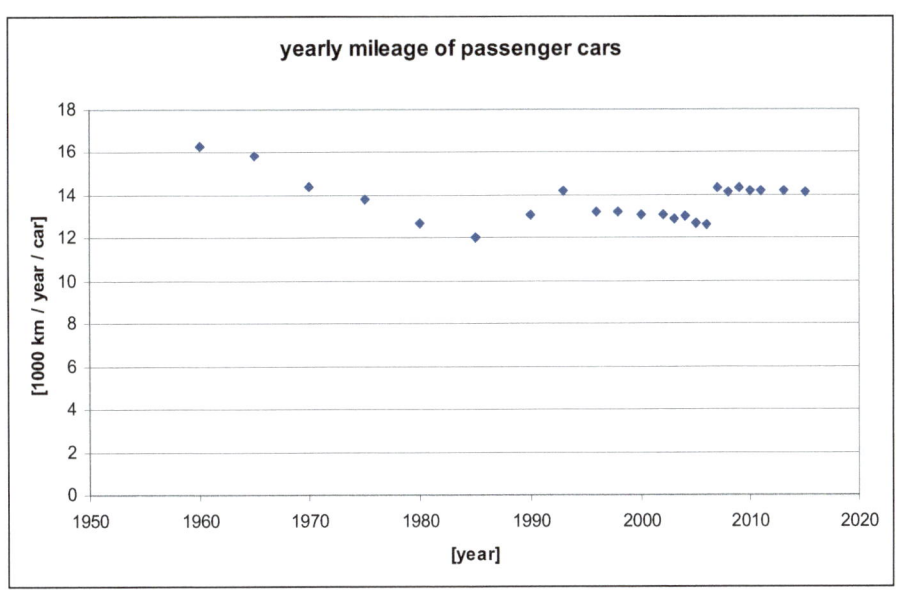

Fig. 13 Stability of the yearly mileage over the last 20 years[1,2,3,4,5]

On the other hand the number of car breakdowns decreased from 23 per 1000 cars / year in 1978 down to less than 3 breakdowns per 1000 cars / year in 1993[6]. Meanwhile, the least defective car is even down to less than one breakdown per 1000 cars / year (BMW X3, year of manufacture 2013).[7]

In other words, the quality, measured in defect driven breakdowns, of passenger cars has increased by a factor of up to 20! And yet the car manufactures' as well as the customers' expectation is to produce and to receive still a better quality.

> **Obviously the people's attitude regarding high quality has changed.**

It is assumed that this change of attitude continues which requires further improvement of product quality in all areas. Also, it has to be kept in mind that the complexity of cars and the number of components have increased, so the real quality improvement per

---

[1] Kunert, U. et al. (2012)
[2] Kalinovska, D. et al. (2005)
[3] T-online (2015)
[4] Kraftfahrtbundesamt (2016)
[5] Steierwald, G., Künne, H.-D. (1994)
[6] ADAC, Grundlagen der Auswertungen in der ADAC Pannenstatistik
[7] ADAC (2016)

component has increased by much more than a factor of 20, and in addition this means for the further increasing complexity of cars, the quality of the components must go up, even if the objective is to keep the current overall quality level of the motor vehicle stable.

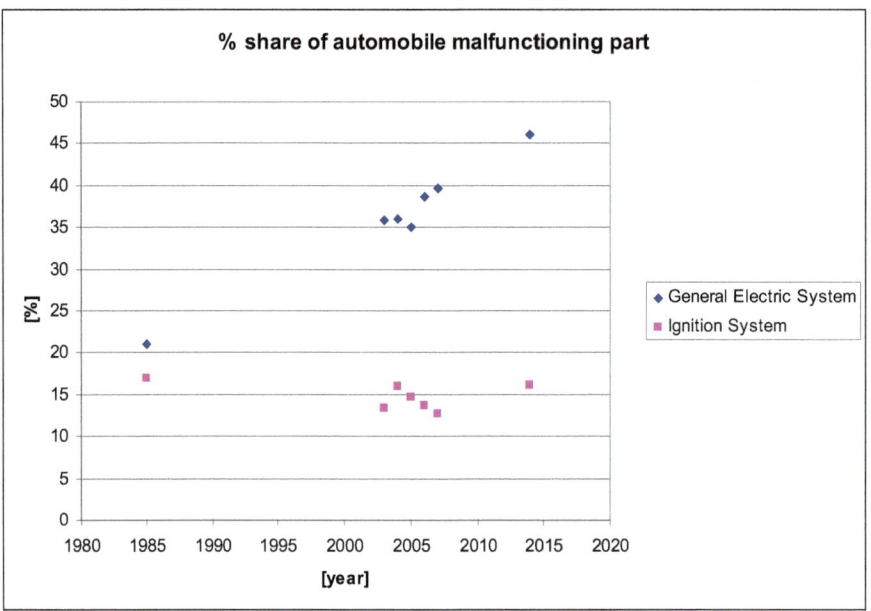

Fig. 14 Share of Electric System based car breakdowns in Germany

When looking into the breakdown statistics of the big automobile associations it becomes obvious that the root cause for automobile breakdown is, to a large extent, found in the general electric system. In addition to that, this share increased significantly over the last 30 years (fig. 14, 15).

However, it is important to mention that the biggest contributor to malfunctioning cars is simply a malfunctioning battery, in particular after six to seven years of use.[1]

Both software and hardware contribute evenly to the automotive electronics fails[2], which requires a low defect approach at both the hardware as well as the software suppliers.

---

[1] ADAC (2016)
[2] Baumann, G., Brost, M. (2008)

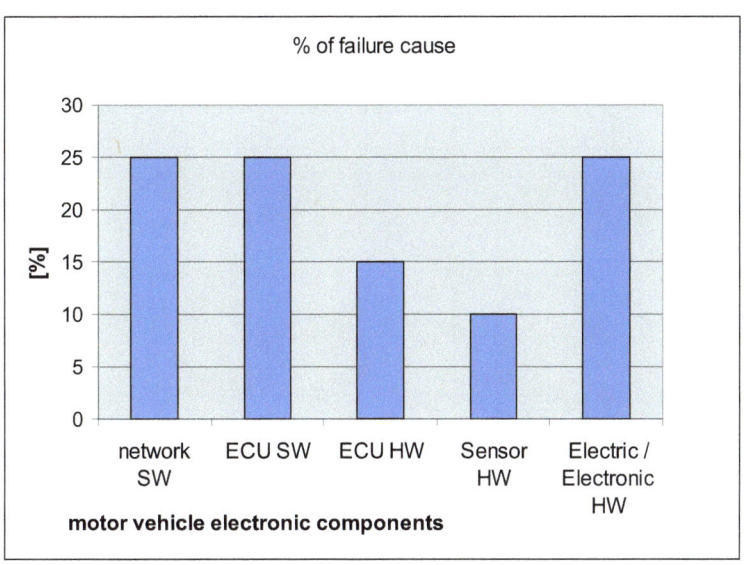

Fig. 15 Percentage of automotive electronic failures

5.1.2. Future Trends

The abovementioned strong increase of the share of the general electric components and systems as a root cause for passenger car breakdown is assumed to be due to the increasing electronic content in the cars. And this trend is expected to continue over the coming years, in particular when thinking about autonomous driving and electrically powered cars with a significantly increased number of electronic devices per car.

More concrete, the number of electronic control units (ECU) per motor vehicle has increased tremendously over the last decades (fig. 16)[1,2], the number of functions executed by one ECU is increasing in addition – today one ECU contains typically 40 - 50 semiconductor devices. A significant growth of the ECU performance measured in million instructions per second and in terms of memory has been observed in the last years (fig. 17). The processor performance required for level 5 autonomous driving is expected to by 100 … 1000 times higher than today's performance.

Those figures correspond to the increasing average semiconductor content per light vehicle in US $ as well as to the share of electronics cost per vehicle. For example, in 1950 the automotive electronics cost was in a range of 1% of the total car cost. Today

---

[1] Baumann, G., Brost, M. (2008)
[2] Polte, T., Aal, A. (2014)

this value has increased to 30 – 35% of the car cost and a further increase up to 50% in 2030 is expected.[1]

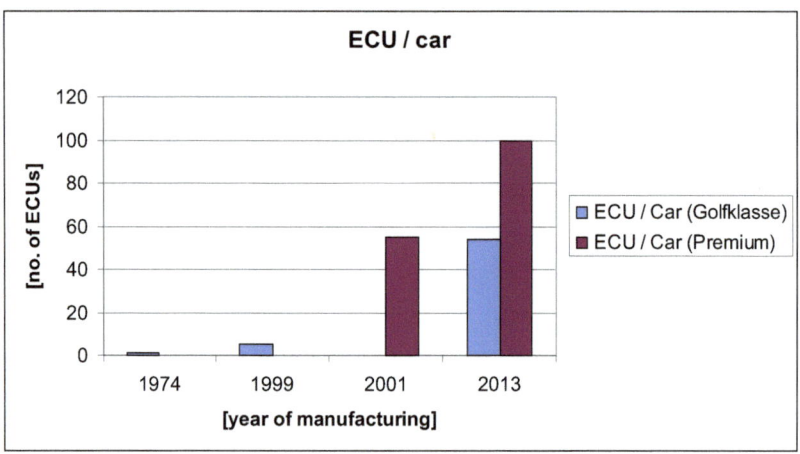

Fig. 16 Electronic Control Units per passenger motor vehicle

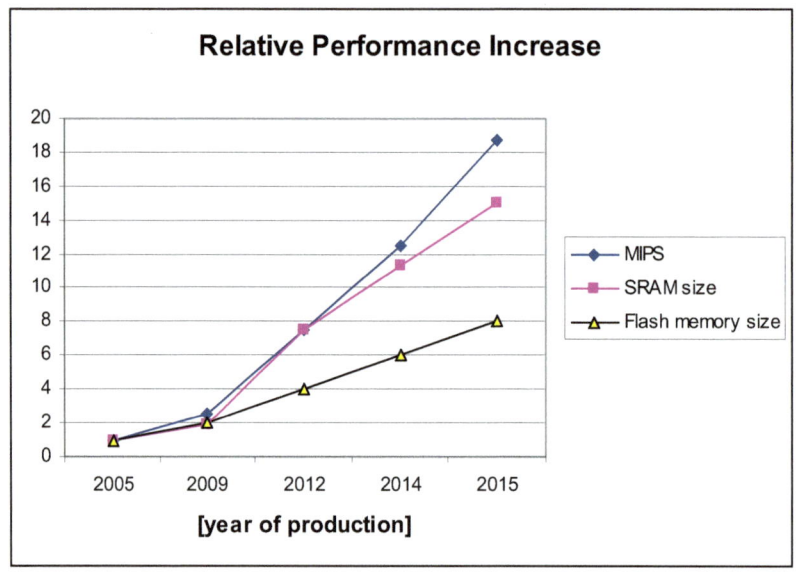

Fig. 17 Relative performance increase of automotive ECUs over 10 years[2]

---

[1] PwC (2013)
[2] PwC (2013)

Hence, assuming a defect rate of 1 dpm per semiconductor unit, the fail rate of an average ECU would be in the range of 40 to 50 dpm. And this, eventually, leads to a electronics hardware based passenger car fail rate of 5000 dpm in today's premium class.

It is obvious that a fail rate of 0.5 %, which equals 5000 dpm is not reasonable to a customer. In addition, considering today's and future safety critical applications, e.g. Anti-lock Breaking System (ABS), Electronic Stability Program (ESP), Airbag electronics or, in general, autonomous driving, defect rates have to be far away from the percentage range. They have to be as low as possible!

Estimation of a future automotive semiconductor fail rate requirements

Let's assume an accepted electronic based passenger car fail rate of 100 dpm in the future. The number of ECUs per car might increase to 200 per car and the number of semiconductor devices might be 200 per ECU, too.

Then simple calculation shows that each electronic control unit must not exceed 0.5 dpm and each semiconductor device must not exceed 0.0025 dpm or 2.5 defects per billion. In addition this has to take into account that the semiconductor devices used become more complex and bigger in size (e.g. the SRAM size grew by a factor of 15 in ten years).

Being aware that evidence for those low defect rates is quite difficult to provide by the manufacturer of components, this seems to be the current or near future level of what Zero Defects means. Grouping of several products may be required to larger production volumes for the calculation of defects per billion.

> The next level of Zero Defects means „defects per billion"

5.2. Low dpm Quality Methods and Tools in the Automotive Semiconductor Industry

Quality tools and methods are the backbone of an agile quality management system. They are often used on a daily basis to improve procedures and product quality and support the implementation of the requirements of the respective quality management system standard.

However, quality tools & methods are typically not precise quantitative methods. Examples for this kind of tools are 8D, Change Management, PDCA, etc., etc.. In contrast to those, FMEA and FTA are examples for quality methods based on figures to express the probability of defect occurrence.

In particular, the FMEA is a widespread quality management method for the proactive investigation of potential failures, their potential root causes and their potential effects on products and processes frequently using quantitative figures for the occurrence of failures.

The following section shows the challenges the FMEA method is facing if applied to concrete sub dpm failure occurrences.

5.2.1. General Tool: Failure Mode and Effects Analysis

The FMEA method is based on a team approach targeting for a systematic analysis of problems which could happen in the future.

In the semiconductor industry the FMEA method is used during the entire phase of product development and manufacturing, here in particular when dealing with product or process changes, e.g.

> ➢ Design FMEA starting in the early phase of product development. The Design FMEA needs to be updated in the different phases of the product development project.

> ➢ Process FMEA starting in the early phase of process development. The process FMEA needs to be updated in the different phases of the process development project. It includes FE processes and BE process as well.

> ➢ Change FMEAs are a mandatory part of the process change process. Depending on the result of the FMEA, i.e. the risk determined with the process change, different risk mitigating actions need to be performed to avoid failures.

The idea behind is to identify potential failures of new products, processes or due to changes in a brainstorming session. For those failures all potential root causes and also all potential failure effects are determined. Each cause-to-effects chain is then assessed regarding

> ➢ Probability of occurrence of the failure (O)
> ➢ Severity of the failure effect (S)

> Detection of the root cause, the failure or the failure effect (D)

Probability, severity and detection are then evaluated using a score from 1 to 10 with 1 being the best case and 10 being the worst case. The risk emerging from these three aspects of a failure is then expressed by the risk priority number (RPN):

$$RPN = O \times S \times D$$

Typically, RPNs equal or larger than 100 require a corrective action in order to prevent this failure from happening. O, S and D figures correspond to qualitative statements like "customer annoyed" or "happens frequently" or to quantitative statements like "50 dpm" or "5 times a week". Both approaches have their advantages and disadvantages. An exemplary occurrence & detection rating scheme which is typical for current in-company standardized tables is shown in table 2.[1,2]

It must be mentioned here, that the determination of an accurate figure for occurrence of specific failures and especially for the escape rate is very difficult. From own experience, the rating is often based on an expert's qualitative knowledge. Frequently the figures are not proven by quantitative assessment, which might lead to major uncertainties regarding the actual occurrence rate at the customer. This uncertainty is well known but is not considered to be a blocking point for the evaluation of this method in a more fundamental way. This is because the main purpose of an FMEA is not the determination of absolute failure probabilities but a ranking of different failure modes according to their respective risk priority number. Experience shows that this works fairly well, so the FMEA is a useful tool to focus resources for failure prevention and quality improvement where they are most needed. In other words, the failure prevention process is made more effective and the same time more efficient.

Also, when assuming the dpm values for failure occurrence and detection from table 2 then the actual failure occurrence at the customer can be calculated by multiplying those figures with each other.

64 out of 100 possible combinations end up with O x D values above 1 dpm at the customer (fig. 18). Assuming more than one hundred line items in the FMEA the probability of reaching a product failure rate in the one digit dpm area is very low. On the other hand, only 14 combinations lead to equal or less than 0.01 dpm. Again, assuming a regular process FMEA in the semiconductor industry to have more than one hundred line items, those figures show, that with the given rating scheme shown it

---

[1] AIAG (2009)
[2] Wittmann, J., Bergholz, W. (2016)

is very difficult to use these risk priority numbers in the sub dpm area on the product level.

Using the above shown figures the "still accepted" occurrence, i.e. the worst case, of failures at the customer can be calculated if a RPN < 100 is assumed to be "allowed" (table 3).

In other words, a Severity = 7 failure, e.g. "customer unsatisfied", could have an occurrence of five dpm at the customer. The overall dpm rate for severity 7 up to 10 would then be 18.33 dpm.

Hence the outcome of these considerations is that the FMEA is a highly valuable quality tool for the assessment of risk in general, but does not provide a complete quantitatively reliable set of data about the occurrence of failures at the customer in the dpm or even in the sub dpm area.

Rating	Occurrence [dpm]	Detection [escape rate dpm]
1	< 0.66	< 10
2	6.66	100
3	66	500
4	500	1000
5	2500	2000
6	12500	5000
7	50000	10000
8	125000	20000
9	333333	50000
10	> 500000	100000

Table 2: Example for a FMEA probability rating for various failure occurrences according to the Automotive Industry Action Group[1]

Therefore, additional methods have to be implemented and in place in order to achieve sub dpm product quality. Several of those methods will be described in the following sections. The focus here, however, is not only to describe the technical method, but also to carve out the philosophy behind.

---

[1] AIAG (2009)

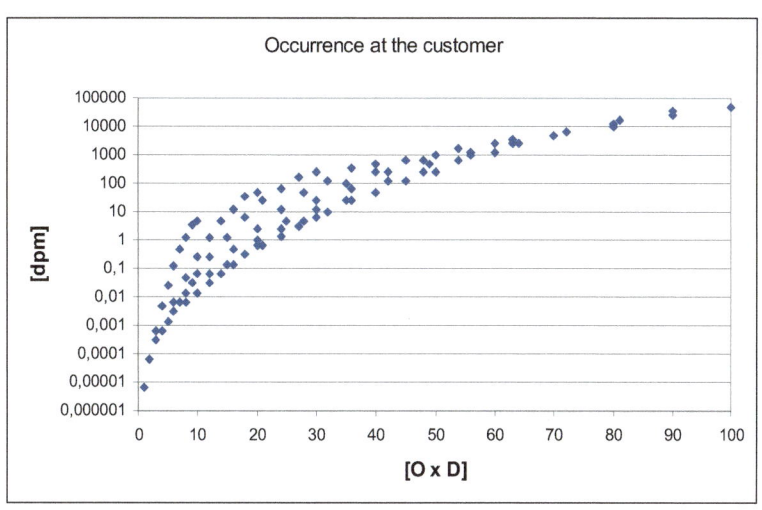

Fig. 18 Failure occurrence at the customer calculated from failure occurrence and detection

Severity	"worst case" occurrence at customer [dpm]
1	33,333.33
2	666.66
3	250
4	50
5	12.5
6	12.5
7	5
8	5
9	5
10	3.33

Table 3 "worst case" occurrence figures at the customer depending on the severity rating and an assumed maximum "allowed" RPN = 100

5.2.2. Low dpm Tools in Design

During the design phase of a product development project all necessary product requirements and functions are transferred to an electrical circuit solution.[1] The outcome of this process is the circuit diagram which is later on transferred to the layout of the integrated circuit.

In this phase, additional units, e.g. additional devices, dummy structures or even functional units, can easily be placed on the future chip or existing units can be arranged in a specific way without any major effort.

Therefore, the device producing companies are required to set up a business process which collects all design related requirements from all affected departments and to transfer those requirements to the design team prior to tape out, i.e. before sending all the data to the mask manufacturer. A few of these methods are introduced here in order to explain the idea behind.

Design for Manufacturability (DfM)

In general, the term "design for manufacturability" implies that the design engineer has to design a product in a way that it can be manufactured at all or at least easier than with a design which does not take potential difficulties in the production process into account. Hence the design engineer needs to know precisely what are the limitations of the manufacturing process as expressed by the known process capabilities for existing processes or the estimated process capabilities for new processes. Otherwise the product cannot be produced or at low yield or high failure rates only. Hence, one interpretation of Design for Manufacturability is:

> "The practice of designing the circuitry so that the part can be more easily manufactured via larger design margins." [2]

Well known examples for the DfM approach can be observed in lithography[3]. The use of standards regarding components as well as production flows improves the manufacturability of a product. On the one hand the manufacturing process can be optimized for a small number of standard components, which means that engineering

---
[1] Wittmann, J., Bergholz, W. (2016)
[2] AEC Q004 (2006), p. 13
[3] Da Silva, M. G. et al. (2002)

resources are not split for many low volume or customized components. On the other hand, the use of standard components enables to produce them in high volume which is required to detect failures occurring on the dpm and sub dpm level. In addition, well designed and reliable standard flows, in particular when combined with a modular production system, are part of an efficient DfM process.

On the more technical side the lithography process, in particular for sub 100 nm technologies, requires resolution enhancement techniques (RET) in order to achieve the desired process result:

- Optical Proximity Correction (OPC)
- Sub Resolution Assist Features (SRAF), also known as Scatter Bars
- Phase Shift Masks (PSM)
- Multiple Patterning Techniques
- ...

When going to smaller feature sizes the pattern on the mask is not transferred to the wafer in the desired way, e.g. corners on the mask result in a rounded shape on the wafer. Additional structures on the mask which do not appear on the wafer in the form of an independent structure (both OPC and SRAF) support the appearance of the target structure on the wafer. This process can be understood in a similar way to how an equalizer enhances the fidelity of music emitted from an audio system.

Phase Shift Masks on the other hand, improve the resolution of the lithography process significantly by exploiting the destructive interference phenomena. Finally multiple patterning techniques came up which split a lithography mask level into several mask steps with somewhat relaxed structures.

All those techniques require adjustments on the mask level and have to be considered by the design engineer. More detailed technical information is available in the literature.[1,2,3,4,5]

However, the design support for the manufacturability of semiconductor manufacturing processes is not limited to lithography only. Other processes, e.g. Chemical Mechanical Polishing (CMP), benefit from measures from the design phase, too.

The purpose of the CMP process is to flatten the surface of the wafer for the lithography process. Unfortunately, the process suffers from weaknesses like dishing and erosion (fig. 19) which need to be avoided for an acceptable process result.

---

[1] Liebmann, L. W. et al. (2001)
[2] Liebmann, L. W. (2003)
[3] Lin, C. W. (2007)
[4] Wittmann, J., Bergholz, W. (2016)
[5] Pan, D. Z. et al. (2013)

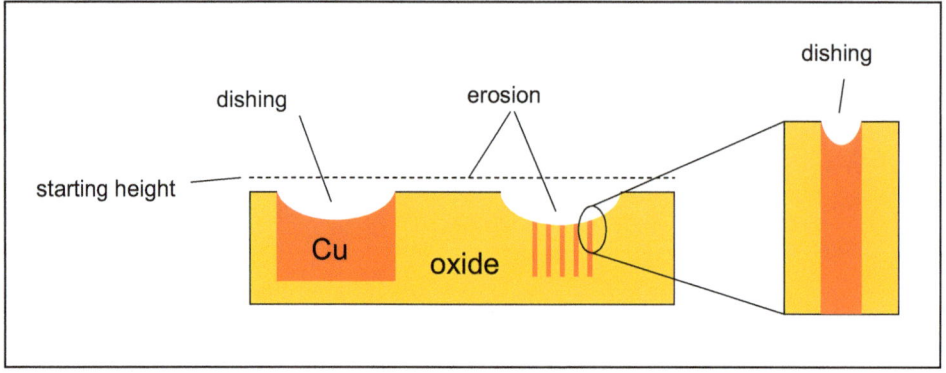

Fig. 19 Dishing and erosion of oxide and Cu structures after a CMP process

In order to achieve a more uniform process result the pattern density on the mask has to be as uniform as possible, too. Therefore areas in the layout with low pattern density for electrical and/or functional reasons are filled with appropriate dummy structures.[1]

Design for Reliability
Different from DfM Design for Reliability keeps the product reliability in the focus. Extensive simulations, reliability testing or field fail analysis are used to identify weaknesses in the product design which eventually lead to a fail of the product under use conditions at the customer. Once identified the design weakness can be overcome by a design change which leads to a more robust product.

A best practice summary regarding Design for Reliability was given by Adamantios Mettas:[2]

"1) Reliability must be designed into products and processes using the best available science-based methods.
2) Knowing how to calculate reliability is important, but knowing how to achieve reliability is equally, if not more, important.
3) Reliability practices must begin early in the design process and must be well integrated into the overall product development cycle."

---

[1] Kahng, A. B., Samadi, K. (2008)
[2] Mettas, A. (2010)

Hence, Design for Reliability has to be managed using a standardized business process. A proposal for a generic process will be given later.

One prominent example for DfR is the use of redundant VIAs. A "vertical interconnect access" (VIA) is the interconnect between two metal layers of an integrated circuit and is susceptible to electro migration effects which eventually lead to a fail. Placing a second, redundant VIA next to the original VIA reduces the risk of fail due to electro migration significantly. Of course a redundant VIA requires additional surface area on the chip and might also lead to circuit timing changes.[1] Both have to be taken care of by the design engineer. An example from photovoltaics (PV) is that over time the number of bus bars (current collection metal strips on PV cells) increased from 2 to at least 3, mainly due to the higher reliability of such designs, at the small cost of a slightly lower efficiency and slightly higher material cost.

Other examples for design parameters, now taken from power electronics modules are:[2]

- Solder joint thickness
- Wire diameter and loop height
- Thickness and shape of a busbar

Changing those parameters directly result in changes of the reliability of the products.

Design for Testability

According to Moore's law, during the last decades the number of components within an integrated circuit double every year resulting in an exponential growth. At the same time the number of metal layers increased from one to ten for the most advanced technology of the respective time. Hence the number of interconnects is increasing exponentially, too.

Due to the fact that the number of I/O pins per integrated circuit does not grow at the same pace, more and more internal nodes become not directly accessible for testing. This leads to a reduction of test coverage, i.e. not every node and not every function of the integrated circuit can be tested directly anymore. Hence the design engineers are asked to come up with methods to increase the test coverage in a more indirect way, for example:

- Use of scan chains
- Insert test points

---

[1] Junping, W. et al. (2014)
[2] Lu, H. (2009)

- Partitioning a large circuit into small blocks for test (physical and electrical partitioning)
- Built In Self Test (BIST): Integrating pattern generation and response evaluation
- …

Frequently these DfT methods require additional chip area which drives cost for the sake of product quality.

Simulation

An important method to evaluate the functionality and the behaviour of integrated circuits early in the development phase is to use simulation software. One popular and well known software package for this task is SPICE, the "Simulation Program with Integrated Circuit Emphasis".

It is used to validate integrated circuits and board level designs and predicts the behaviour of the circuit early in the design phase. Being aware that the manufacturing processes for the fabrication of integrated circuits suffer from process variations and also being aware that the devices and metal lines etc. suffer from aging effects in the course of the use time the simulation approach needs to be used in a dynamic way:

- Process Variation

    "For robust designs the influence of process variations has to be considered during circuit simulation."[1] This influence can be included in an existing SPICE model in two ways: either the worst cases of single device performance are used to model the circuit behaviour or the variation of the device performance is based on a Monte Carlo method. The first approach might lead to very pessimistic results, whereas the Monte Carlo method might need a high number of runs to generate a reliable result.

- Reliability Prediction[2,3]

    The use of integrated circuits leads to electrical or temperature stress of the single devices. This stress might lead to a change of the device behaviour over time, i.e. the devices are aging.
    Inserting the device behaviour into a SPICE model can then be used to predict the integrated circuit reliability performance.

---

[1] Rappitsch, G. et al. (2004)
[2] Wang, W. et al. (2007)
[3] Li, X., Qin, J., Bernstein, J. B. (2008)

Both the impact of process variations and simulated circuit reliability can be used to identify weaknesses of the design way before the mask production starts.

Design for Quality Process & Learning for Zero Defect
Design for Quality, in general comprises the above described approaches together with other quality enhancing methods in the design phase, e.g. Design for Analysis or Design for Diagnostics. From a Quality Management point of view all these methods have to be used in a systematic and consistent way, i.e. they have to be mandatory and they have to be implemented in procedures and business processes.

Fig. 20 shows a generic process describing the return loops from qualification and customer fails back to the design process. It is obvious that product fails may be caused by manufacturing problems or insufficient test coverage, too, which is why they are also indicated in the process.

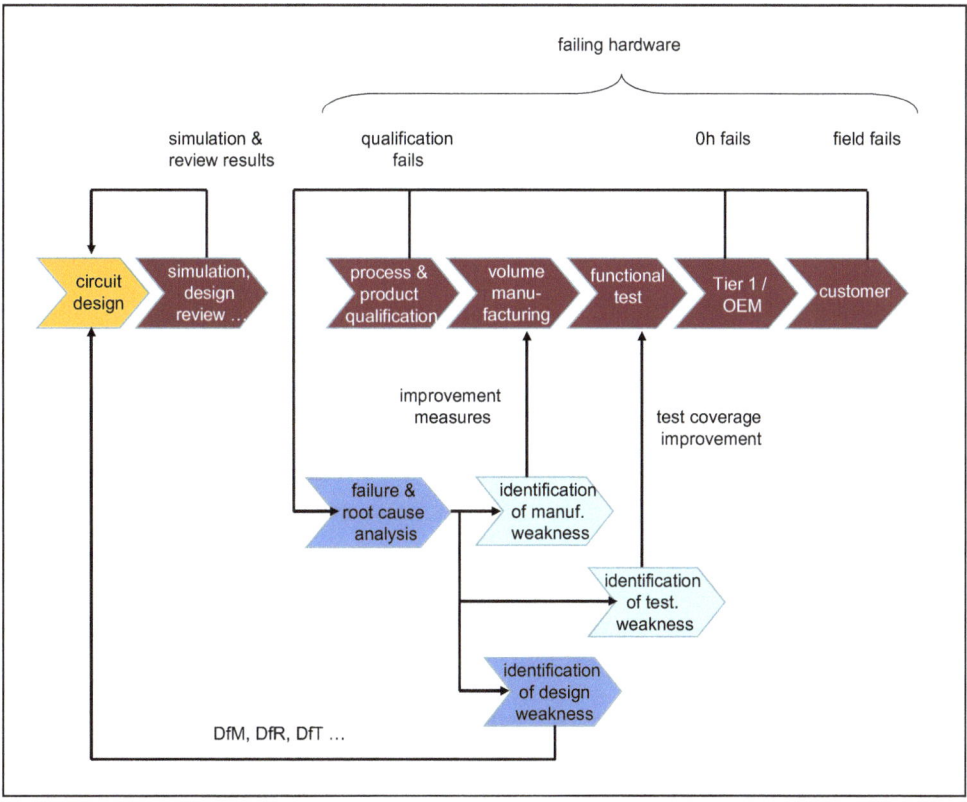

Fig. 20 Generic process starting from the design phase with software and hardware based feedback loops regarding design issues

The learning for the design phase is clearly that the aim for zero design related defects requires a holistic view throughout the entire product lifetime and throughout the entire company. This Zero Defect Design process has to be part of a company wide Zero Defects process and must be supported by a key performance indicator system with regular reporting to the management.

5.2.3. Low dpm Tools in Development, Qualification and Ramp Up

Development

It goes without saying that the development process has to comply with typical quality management system standard requirements like phase wise development, milestone reviews and release gates. The use of basic quality management rules is as well a must, but not sufficient to reach dpm or sub dpm failure rates.

However, the milestone checklists and review lists in technology as well as product (integrated circuit) and package development have to include Zero Defects related check points, e.g.

- Were all innovations and changes in design / process / test / mission profile / customer expectation etc. as compared to previous products identified and listed?
- Has a cross functional team of design / development / production / reliability / test engineers discussed the above mentioned list regarding potential manufacturing / reliability / test issues coming with the new product / technology / package?
- Were all items on the list assessed regarding potential increase in defect rates and/or reliability, manufacturability, testability and, if required, was a corrective action in design or manufacturing / test etc. initiated?
- Were all findings and assessments documented in the project documentation?

Qualification

Qualification is "The entire process by which products or production technologies are obtained, examined and tested, and then identified as qualified"[1]. In case of semiconductor devices this includes stressing and testing the product according to the respective norm or customer requirements (stress test based qualification).

During the qualification phase the product is exposed to various stress conditions, e.g. temperature, humidity, voltage etc., which exceed the use conditions considerably, over

---

[1] ZVEI (2007)

a period of typically 1000 hours, or cycles in the case of temperature cycling. If required, this may be extended to 1500 hours or 2000 hours, or even more.

However, typical sample sizes are three lots with 77 pieces per lot per stress condition, which means a maximum of 1% failures with 90% confidence level in case of zero fails in the test. In other words, standard qualification processes cannot reveal any information about expected dpm levels but provide information about product reliability in general on a moderate level. In other words, disaster can be prevented but test for reliability on a dpm level is not achieved.

Robustness Validation

One method to learn more about the actual reliability performance is called robustness validation. Robustness validation determines the point of failure depending on defined parameters, e.g. by significantly extending the stress test time intentionally beyond the standard requirements or customer requirements: the parts are stressed until they fail (test to fail). In addition to stress time other parameters could be supply voltage, number of temperature cycles or temperature range of cycles. It could include product specific properties like the number of VIAs, too.

Fig. 21 shows the two dimensional view of application requirements placed well within the component capabilities with two parameters A and B. The distance between the edge of the application requirement to the edge of the component capability is the robustness margin.

Of course, the extension of the stress duration (e.g. stress time, number of cycles) sooner or later leads to the fail of the product. Plotting the cumulated number of fails under stress conditions in the probability plot and applying the calculated and/or experimentally acceleration factor to it leads to the fail distribution under use conditions (fig. 22). From there the dpm rate can be estimated, too, provided the failure mechanism has not changed.

Learning

The Zero Defects philosophy does not only aim for a successful development process ending up in a successful product qualification according to the quality standard requirements, but seeks the complete understanding of the product under use conditions, the failure modes and mechanisms including their interactions and the respective acceleration model.

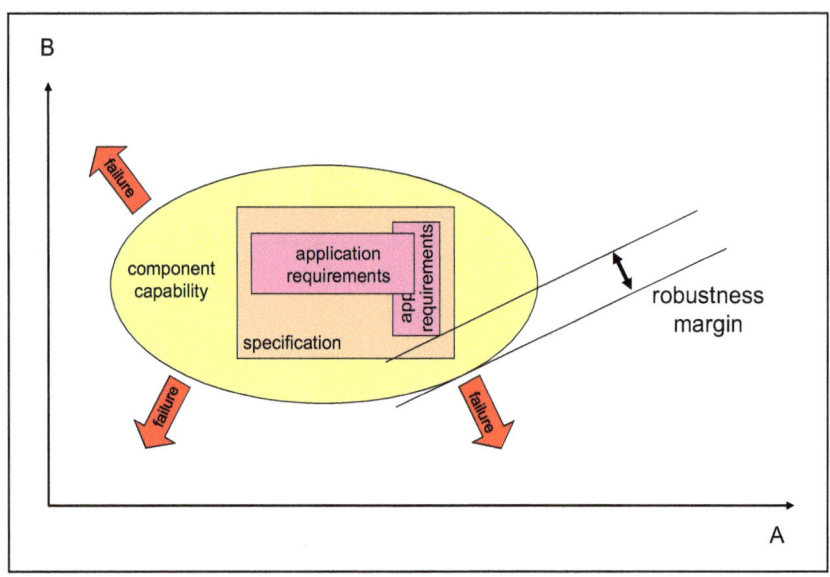

Fig. 21 Visualization of the margin between the edge of the application requirements and the actual product / component capability[1]

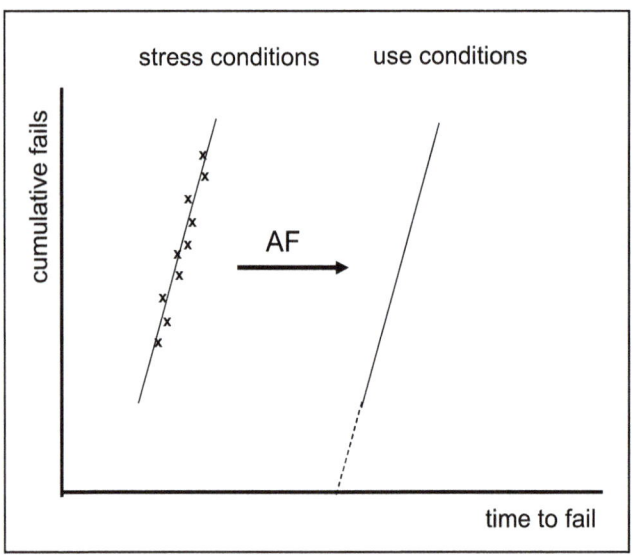

Fig. 22 Schematic extrapolation of stress-to-fail data to operating time to fail (AF acceleration factor) under use conditions

---

[1] ZVEI (2013)

However, a successful qualification is not the end of the process – it's just the beginning of the process to implement low dpm in manufacturing!

5.2.4. Low dpm Tools in Manufacturing

Process Control

Independent from Zero Defects process control activities are a fundamental element of a stable, high yielding semiconductor manufacturing process. Hence they are the basis for low defect figures, but they are not sufficient to achieve them. Typical methods are[1]

- machine and process capability studies
- gauge capability studies
- inline process control, e.g. control of thickness, uniformity, alignment, critical dimension…
- use of process control monitoring with separate test structures for physical (inline) and electrical parameters
- electrical process control, e.g. voltage parameters, frequencies, …
- statistical process control applied to both inline and electrical parameters
- advanced process control
- functional test on wafer, chip and module level
- outlier detection methods applied to the above mentioned tools
- change control
- …

Those methods in combination with a general defect monitoring and defect reduction programs keep the process stable and the defect density low.

Safe Launch

Originating from the development process which ended with a successful qualification the product is now allowed to be produced and delivered to the customer. Unfortunately, the production in high volume frequently reveals instabilities or minor issues of the manufacturing process or the product which have not been identified during development, but lead to a higher failure rate than acceptable.

In order to avoid the delivery of products with higher failure rate to the customer at the beginning of the production phase safe launch measures are implemented in addition to the regular quality control measures like increased inline sampling rates, tighter limits for inline process control as well as electrical and functional tests, additional parameters to be controlled or additional reliability monitoring.

---

[1] Wittmann, J., Bergholz, W. (2016)

Only after the verification of the process and product stability (including test coverage) those additional safe launch measures will be discarded.

Manufacturing standards

In general materials, processes, equipment, geometry etc. should be standardized as much as possible. Not making use of this principle, e.g.

- use of different material suppliers for the same application
- use of different kinds of equipment for the same process technology
- use of different process flows for the same kind of product

such procedures naturally lead to an increase of variability in the production flow. Often there are good reasons for this kind of variability. A second source for materials might be mandatory, a customer insists on the usual process flow or it makes no sense to buy outdated equipment for the extension of the manufacturing capacity might all be the reason for high variability in production.

From a Zero Defects point of view however this is counterproductive. The focus on one kind of material to be processed in one kind of standardized equipment with one kind of process flow would allow manufacturers and designers to optimize process and design to those given boundary conditions.

Reliability Control

In addition to fails at the beginning of use or production, measured in dpm, also the fail rate later in time is very much relevant for the Zero Defect concept. Those reliability related defects are typically not measured in dpm, even though some of the supplier rating procedures describe it that way, but in FIT (failure in time), i.e. fails per 10^9 hours of use.

Three main methods are presented here which provide information about the expected reliability performance of the produced parts:

- Wafer Level Reliability (WLR) & fast Wafer Level Reliability (fWLR)
- Burn In (BI)
- Product Reliability Monitoring

For WLR & fWLR special test structures are placed on the wafer, e.g. in the sawing line, which can be stressed by applying higher current and voltage conditions to them. Hence these methods, assumed the failure mechanisms are very well known, can be used on wafer level in a short period of time. So the benefit here is a very short feedback loop regarding potential reliability issues.

Burn In (BI) is frequently used to artificially age the parts by applying BI stress conditions to all of them. Having the bathtub curve in mind, this means, that weak parts, e.g. with extrinsic defects, are screened and the remaining parts with low defect level may be delivered to the customer. A reduction of the BI yield hints towards a hidden reliability problem and should be used as reliability performance indicator.

Eventually also finished parts can be used for product reliability monitoring (PRM). This kind of monitoring is often done accompanying to production and delivery and is very time consuming compared to other methods, i.e. includes all packaging activities and stress of about 1000 hours. In addition the amount of parts to be stressed is typically comparable to the amount of parts used for qualification. Hence, PRM is not able to provide information about the dpm level but is more a disaster check.

5.2.5. Low dpm Tools in Test

Functional tests are required in order to screen out non-functional parts. Those tests are implemented on wafer and on product level in order to confirm the functionality of the dies. Tests on wafer level are typically more productive and cheaper, but do not deliver all the information required. Some tests, e.g. high frequency tests, require the entire packaged product to be efficient.

In addition to the functional test, integrated circuits have to undergo the parametric tests, too. DC parametric tests measure electrical characteristics under steady-state conditions such as voltage, current, resistance and power consumption. On the other hand, AC parametric tests are applied to measure electrical characteristics like impedance or dynamic resistance. The product may be functionally good, but fails to meet the specification of an electrical parameter.

All those tests and test results are subject to intense analysis. Both systematic fails, e.g. due to photo mask defects, and random fails, e.g. due to particle contamination, are analyzed and measures are initiated in order to reduce or eliminate their impact on the product yield.

All these activities are typical activities in semiconductor manufacturing with no special emphasis on Zero Defects. The following examples, however, should be applied, if targeting for Zero Defects.

Statistical Yield Limit[1]

"Reliability defects are proportional to yield defects (typically 1 % - 2 %)"[2]. In addition to cost and productivity reasons this statement emphasizes the importance of a high yielding process for a reliable product, i.e. low functional yield indicates lower reliability, too. Therefore, low yielding lots (or wafers) are subject to scrap in order to avoid delivery of low reliability parts to the customer.

The identification of low yielding lots is based on the known typical lot to lot yield variation. Once a lot is outside the usual variation, e.g. the lot yield is outside a 4 σ range around the mean yield, it needs to be scrapped if yield is low, or should be analyzed, if yield is high.

In order to make this more clear: a low yielding lot, say with 50% yield, is then scrapped because it might suffer from reliability relevant defects, i.e. in this case also the functionally good half of the lot is scrapped.

Good Die and Bad Bin in a Bad Cluster

The nature of the semiconductor manufacturing process leads to a random or clustered distribution of defects across the wafer surface. In addition, as mentioned above, a defect may not lead to a functional fail, but may lead to a fail later in time.

Assuming the functional test shows a cluster of failing dies with one or two good dies within this cluster, then there is a certain probability that those good dies are functional, but are affected by a cluster of defects which eventually lead to a reliability issue. Hence the Zero Defect concept requires to scrap those good dies, too (Good Die in a Bad Cluster, fig. 23).

Similarly, the edge of a defect cluster is not clearly defined. Good dies neighbouring a cluster of failing dies in the functional test are considered potential reliability issues and are scrapped for this reason, too (Bad Bin in a Bad Cluster).

Again, the idea behind is to discard parts which are good according to the test results but might be bad.

Part Average Testing (PAT)[3]

The values of the parameters tested in parametric testing are typically not static, but show a spread, e.g. according to the normal distribution. Thus, again, the mean or median value and a defined range around them, e.g. a 6 σ range, is then considered the normal case and outliers, i.e. parts with values outside this range, are then

---

[1] AEC - Q002 (2012)
[2] ON Semiconductor (2015)
[3] AEC - Q001 (2011)

considered potentially bad. They will be scrapped, even though they are well within the given specification.

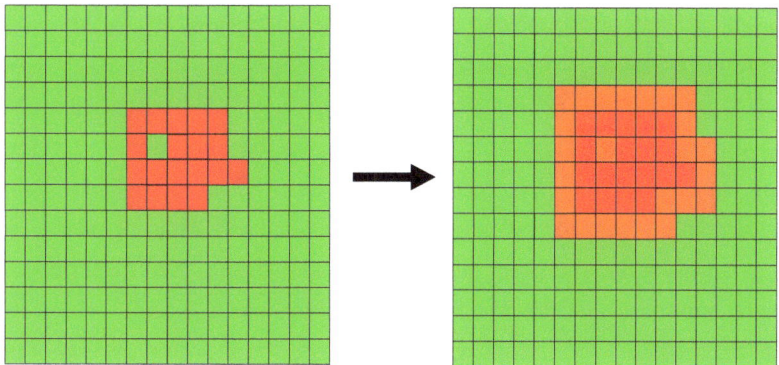

Fig. 23 Good Die and Bad Bin in a Bad Cluster (red: originally bad dies, orange: in addition considered bad)[1]

Screen, e.g. Iddq leakage test

During the quiescent status a defect free integrated circuit is characterized by a very low consumption of electric current. Additional defect driven paths for the electric current, e.g. gate oxide shorts or metal line shorts, leads to a considerable increase of electric current consumption even in the quiescent status due to additional leakage currents. This behaviour is used to identify parts with so far undetected defects and to discard them. This is a very simple test and should be performed at the beginning of the test sequence to reduce test cost. Unfortunately with decreasing feature sizes, i.e. with scaling technology, the background currents become higher which makes it more difficult to identify bad parts.[2]

5.3. Learning and Consequences

5.3.1. Learning

As stated earlier, automotive product quality has been improving significantly by at least a factor of 20 during the last decades. Due to the continuously increasing electronic content of cars and with increasing functionality and calculation power of the electronic control units and semiconductor based sensors and the customer's expectation of higher quality cars, the defect rate of automotive semiconductors has to continue to be reduced from the dpm level down into the dpb level.

---

[1] Wittmann, J., Bergholz, W. (2016)
[2] Nahar, A. et al. (2009)

All standard quality tools, such as Quality Function Deploymen (QFD), 8D, Change Management, Deviation Management, FMEA etc. are a must to achieve high quality products. However, they are not sufficient to achieve dpm and sub dpm product quality, which is why additional, mostly technical methods have to be used, too.

A similar situation is observed in manufacturing: literally all known methods of process control have to be used. Further reduction of the dpm rate can be achieved by applying standards (material, equipment, process flow) as much as possible and to increase process and product control activities in the early manufacturing phase. Reliability indicators, e.g. wafer level reliability results and BI yield, help to identify potential reliability problems prior to shipping the hardware to the customers

In the test area the Zero Defect concept requires testing and methods beyond simple yield improvement and cost reduction. Only extensive usage of mathematical and statistical methods, e.g. to identify outliers, and consequent scrapping of potentially bad parts lead to dpm and sub dpm product quality.

5.3.2. Consequences & Proposed ZD Business Process

Overall a holistic concept of Zero Defects way beyond the ISO9001 standard procedures and throughout the entire company is required. Good examples for this approach are, besides others, Design for Manufacturing, Design for Reliability and Design for Test. Departmental thinking is not successful – an integrated approach with short and long feedback loops from technical experts in other departments back to the Quality Management – or Zero Defects - department is essential to achieve low defect rates.

The following items are considered to be some of the key drivers for further improvements of the Zero Defects concept:

- Continuous further development of technical methods for Zero Defects in all relevant departments
- Continuous further development of quality management methods for Zero Defects
- Set up of a company wide Zero Defects process
- Implementation of a Zero Defect organization, e.g. within the Quality Management department, with empowerment for Zero Defects

A generic process is shown in fig. 24. It includes short and long feedback loops and an overall ZD organization monitoring and controlling the entire ZD process. The main customer related elements of the process shall be described in the following section.

Early detailed information about customer's requirements

Prerequisite for achieving low defect rates is a close cooperation with the customer on a technical level. Of course, the customer should provide detailed requirements ahead of time and the part manufacturer has to receive and understand those requirements and transfer them into the respective product parameter settings. This should be done in a way that for each customer requirement one or more product parameters are identified or defined, which, together, fully support the realization of the requirement.

It is also essential for achieving sub dpm product quality that the part manufacturer (e.g. Tier 2) understands not only his direct customer (Tier 1, sub-system manufacturer), i.e. his product, but also the OEM's product, i.e. the entire system. Understanding the operational modes of the system enables the part manufacturer to anticipate operations critical conditions and to initiate corrective actions in a proactive manner.

In addition, a detailed knowledge of the working conditions, described in the mission profile, allows the part manufacturer to strengthen potentially weaker areas regarding reliability or performance by design. In this case knowledge of the final customer's habits needs to be compiled, too.

It goes without saying that this kind of system engineering requires engineering resources way beyond "side job" efforts. A separate organization similar to a competitor analysis group could be an option for this.

Management of External Fails

Unfortunately, "zero defects" is not possible, which means, that products fail at the customer side. From a Tier 2 point of view several kinds of fails have to be distinguished:

(1) 0-mileage fails at Tier 1: production fails (incl. incoming inspection)
(2) 0-mileage fails at OEM in production (incl. incoming inspection)
(3) 0-mileage fails at final customer
(4) Field fails at final customer within warranty
(5) Field fails at final customer after the warranty period
(6) End of life fails

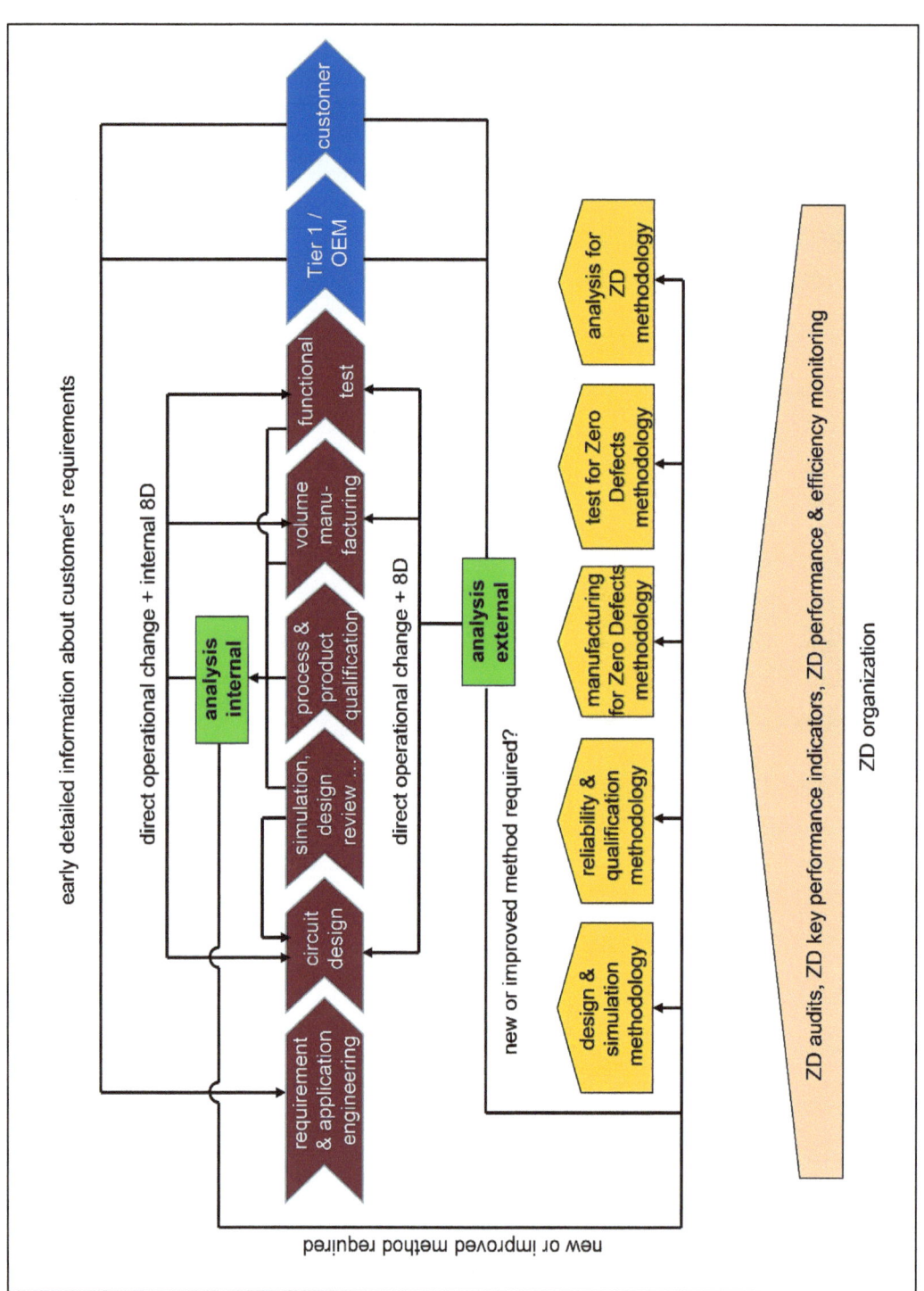

Fig. 24 Generic Zero Defects process

Besides differences regarding the severity of the fail related to compensation etc., these cases represent different levels of complexity regarding prevention or detection of the fail prior to delivery and/or entirely different failure mechanisms. In addition, it may be difficult or even impossible to receive field fails after the warranty period for analysis.
After receiving the failing part with failure, failure mode and application description all cases require a bottom line standard procedure including retest, simulation, physical analysis etc. with details depending on the particular case. Eventually, after solving the problem, the 8D report needs to be sent to the customer and the root cause needs to be inserted into an 8D database.

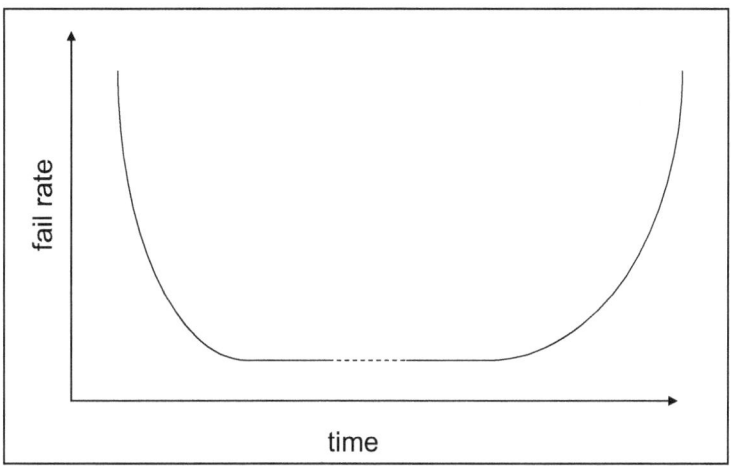

Fig. 25 Bathtub curve representing (operational) time dependent fail rates[1]

Except for parts produced at the same period of time this root cause must not be in the data base already. If it is, then the 8D process has to undergo a review.
Typical fail rates of the above mentioned different cases follow the so called bathtub curve (fig. 25).

The early infant mortality period with decreasing failure rates is dominated by process and design weaknesses, material defects, problems during assembly etc.[2]. In addition the test coverage seems not to be sufficient to catch those failing parts. Preventive measure might be stress tests in order to sort out weak parts before sending them to the customer.

---

[1] Wilkins, J. D. (2017)
[2] Wilkins, J. D. (2017)

Cases 1 to 3 represent this kind of fails which occur at the very beginning of the product life. Ageing and wear out has not taken place yet and several functional tests were performed prior to delivery:

Case 1: 0-mileage fails at Tier 1
In this case the manufacturer's final test failed to detect the defective part. However, during production, i.e. implementation into the subsystem or at test of the subsystem, the part turns out to be malfunctioning. Hence the root cause analysis should include system influences and part and system manufacturer may have to cooperate to solve the problem.

Case 2: 0-mileage fails at OEM
In this case the manufacturer's final test, Tier 1's intermediate tests during production and final subsystem test failed to detect the defective part. However, during production, i.e. implementation into the overall system or at test of the system, the part turns out to be malfunctioning. The question here might be what is the additional impact of the overall system on the part behaviour. Are there additional or alternate operational modes which are not known to the part and/or the sub-system manufacturer? Maybe the sub-system manufacturer is not aware of all possible system influences on the single part. In this case, when the sub-system manufacturer was not successful to solve the problem, it might make sense to initiate a direct contact between part manufacturer and OEM. Only then the complete set of data and information can be exchanged without unintentional loss of information. Of course all parties should be involved.

Case 3: 0-mileage fails at final customer
In this case all tests during and at the end of production failed to identify a weak part. The part turns out to be malfunctioning at the final customer only. Although some OEMs consider very low mileage failures still as 0 mileage failure, this defect is not considered a real lifetime issue.
The question here might be what has changed between final test and beginning of operation? Was the mission profile incomplete? Or is there anyhow an unknown fast working reliability issue leading to part malfunctioning? Similar to case 2 cooperation of all manufacturing parties is recommended.

The infant mortality period is followed by the normal life period with a low constant failure rate. "Normal life failures are normally considered to be random cases of 'stress exceeding strength' "[1]. However, many fails in the normal life period are actually part of the tail of the early fails. "Failures from infant mortality defects get spread out so much that they appear to be approximately random in time"[2]. Eventually the combination of remaining early fails and random fails, e.g. soft error rates, result in the "normal life failure rate". Cases 4 and 5 are part of the normal life period. Hence the reduction of the early fail rate leads to a reduction of the normal life fail rate.

In the end the products will fail due to wear out (case 6). If this occurs only after a defined or expected life time then the customer is most likely willing to accept it. The wear out phase must not start before case 5.

All of the fails returning from the customer have to analyzed and, for cases 1 to 5, the root cause has to be eliminated or effective screening methods have to be installed. Analysis of case 6 fails should be used to gain a better understanding of the fail mechanisms.

ZD Process Monitoring

In order to meet the functional safety requirements of ISO26262 or the more formal requirements of IATF16949 companies frequently have dedicated groups within the overall quality management organization taking care of the specific requirements.

The same way it would make sense to set up a Zero Defect organization with solely focus on Zero Defects. Its main tasks would be for instance

- Set up and monitor ZD Key Performance Indicators, e.g.
    - dpm level per product, technology, business group, production line
    - FIT level per product, technology, business group, production line
    - Number of incidents
    - Number of 8D reports
    - Burn In yield

- Provide 8D Support
    - Heading or contributing to current 8Ds
    - Investigate 8D reports with same or similar root cause and identify more general root causes

- Implement ZD Audits

---

[1] Wilkins, J. D. (2017)
[2] Wilkins, J. D. (2002)

- Audit and monitor compliance the ZD rules and procedures in all areas (development, procurement, production, quality management, ….)

> Identify and implement methods, e.g.
- Identify and implement new methods of defect reduction, prevention and detection in all areas in a proactive way

Being aware that defect generation, reduction, detection etc. is a very wide field of knowledge and that it requires detailed and in depth knowledge in each of these areas of expertise the ZD organization has to have access to the respective experts in the specialty departments. In terms of ZD they report to the central ZD organization. At the same time these experts are the ZD representatives in their department.

6. Summary and Conclusion

Historical View and Criticism
From the beginning in the 1960s the term "Zero Defect" did not mean that the products are perfect, but always meant the way of preventing defects. At that time many companies – primarily in the defense and aerospace industry - started Zero Defects programs in order to achieve lower defect rates and many of them were very successful with it.

After several years, however, the importance of Zero Defect programs declined as the majority of Zero Defects related activities became ordinary activities in companies striving for high quality products. Separate special ZD programs were not required anymore.

Despite being very popular Zero Defects early on was subject of major criticism. On the one hand the requirement of reducing defects to zero was considered to be demotivating employees rather than motivating them. On the other hand, ZD seems to be impossible to be achieved at reasonable cost. Early cost based criticism of Zero Defects is assumed to be mainly caused by the traditional view on quality cost.

Later, starting in the 1980s, the automotive and aviation industry more and more focused on the production of defect free or, at least, low defect products. At this point in time the number of worldwide publications about Zero Defects started to grow significantly. A quick search for "Zero Defects" on Google reveals, that 35% of the hits in an industrial environment were from companies in the automotive sector, followed by "Aerospace & defense" and "medical, health & life sciences, in descending order.

Industry point of view

Today, in particular in the automotive industry, Zero Defects means the strategy to achieve defect rates at or below one defective part out of one million parts. Hence, up to now a dpm quality level has been considered Zero Defects and below. In the coming years, and this has already begun, expectations move further down to the "defects per billion" level. This means that single events are considered quality issues – in particular when considering the production volume which is required to prove the dpb quality level.

Driven by the technical progress the understanding of Zero Defect is a continuously "moving target" towards lower defect rates at reasonable cost. This can be expressed by applying a logarithmic scale in the cost of quality model.

In order to gain a more detailed understanding of the actually accepted product failure rates and what means Zero Defects in the industry the companies' supplier rating system was examined.

- the majority of documents with clearly defined dpm targets for their suppliers originate from TS 16949 certified companies
- the definition of the meaning of "defects per million" is often not so clear from the procedures
- the time period, i.e. the delivery volume, for the calculation of the dpm level varies from company to company
- many companies talk about Zero Defects requirements, but accept a failure rate on a percentage level
- the automotive industry demands the lowest dpm rates from their suppliers
- all companies allowing a maximum of 200 dpm for the "worst" class supplier rating are TS 16949 certified

Steps towards Zero Defects

Compliance to ISO9001 or IATF16949 and the use of "standard" quality tools like FMEA, FTA, QFD etc. etc. are mandatory, but not sufficient to achieve product quality on the dpm – or in the future – on the dpb level.

On the first level Zero Defects tools and methods have to be implemented all across the company: development & design, procurement, production…. These tools must not be separate tools, but need to be linked together in a more holistic way. Design related tools like "Design for Manufacturability", "Design for Reliability" or "Design for Testability" show clearly what a Zero Defects philosophy means: weaknesses in one area have to be compensated by measures in other areas in order to achieve the overall target. In addition, feedback loops from the customer or other sources back to the origin of failures have to be implemented and used continuously.

The company's mindset has to be adjusted to "quality first", i.e. to avoid any potential fail at the customer by the right measure. For instance in the qualification phase robustness validation is used to identify potential weaknesses and in the early volume phase so called safe launch measures have to be implemented to avoid problems often occurring in the early production phase. This approach also includes tools in the test area, where functionally good parts are scrapped, because they MIGHT be bad, i.e. "good die in a bad cluster", because the neighbouring dies are functionally bad.

The next level of Zero Defect would be to institutionalize the ZD approach. For instance, this would include the definition and monitoring of ZD key performance indicators across the entire company, follow up on fails with similar or same root cause and ZD audits.

Hence, despite the fact that many companies obviously have not yet reached today's acknowledged ZD level (dpm), Zero Defects is very much alive in quality sensitive branches and the absolute level of Zero Defects continues to be a moving target.

7. References

ADAC, "ADAC Pannenstatistik 2016", 2016
https://www.adac.de/infotestrat/unfall-schaeden-und-panne/pannenstatistik/
[2017/02/04]

ADAC, „Grundlagen der Auswertungen in der ADAC Pannenstatistik"
https://www.adac.de/infotestrat/unfall-schaeden-und-panne/pannenstatistik/pannenstatistik_methodik_hintergrund.aspx?ComponentId=259919&SourcePageId=259736
[2017/02/04]

AEC Automotive Electronics Council, Component Technical Committee, "Guideline for part average testing", AEC - Q001 Rev-D, December 9, 2011

AEC Automotive Electronics Council, Component Technical Committee, "Guidelines for statistical yield analysis", AEC - Q002 Rev B, January 12, 2012

AGM Automotive, "AGM Supplier Quality Manual, QF-SQR001 / Rev. Date: 12-5-14 / Rev. Level: D", 2015

http://www.agmautomotive.com/wp-content/uploads/AGM%20Supplier%20Quality%20Requirements%20Manual%2012-5-2014.pdf
[2017/02/18]

AIAG, "AIAG FMEA-Ranking-Tables", 2009
https://elsmar.com/pdf_files/FMEA%20and%20Reliability%20Analysis/AIAG%20FMEA-Ranking-Tables.pdf
[2017/03/05]

Alex Products, Inc., "Supplier Quality Manual Revision I", 2015
http://www.alexproducts.cc/pdf/qualitymanual.pdf
[2017/02/16]

Alfmeier Präzision AG, „Leitfaden zur Lieferantenbewertung"
http://www.alfmeier.de/fileadmin/user_upload/PDF_Uploads/Downloads/Leitfaden_Lieferantenbewertung_Alfmeier.pdf
[2017/02/16]

Amway, "Supplier Quality Requirements Manual", 2016
http://supplier.amway.com/sites/supplierportal/quality/qualityreview/Online%20Documents/Supplier%20Quality%20Requirements%20Manual.pdf
[2017/02/16]

Ardianto, M. N., "Zero Defects And Its Relevance In Modern Industries", mna Quality Consulting, 2014
http://www.monika-ardianto.com/zero-defects-in-modern-industries/
[2017/01/28]

AUSTRALIAN ARROW Pty. Ltd., "Supplier Rating Scheme", QADCP-009 Issue J : 26/08/2013
http://www.australianarrow.com.au/purchasingpdfs/004_Supplier_Rating_Scheme.pdf
[2017/03/02]

Baumann, G., Brost, M., "Testverfahren für Elektronik und Embedded Software in der Automobilentwicklung", Forschungsinstitut für Kraftfahrwesen und Fahrzeugmotoren Stuttgart, Workshop MBEES Dagstuhl, 2008

Boeing, "Supplier Performance Measurement"
http://www.boeingsuppliers.com/prefsup.pdf
[2017/02/16]

BorgWarner Inc., "BorgWarner Supplier Manual", 2015
https://www.borgwarner.com/docs/default-source/default-document-library/borgwarner-supplier-manual.pdf?sfvrsn=10
[2017/02/18]

Brueser GmbH, „Aufschlüsselung der Lieferantenbewertung der Paul Brueser GmbH", 2008
http://brueser-gmbh.de/download-dokumente/bruesergmbh-aufschluesselung-der-lieferantenbewertung.pdf
[2017/02/16]

Capsonic, "Capsonic Automotive and Aerospace Supplier Manual", 2014
http://www.capsonic.com/CA2/wp-content/uploads/2016/10/Capsonic-Supplier-Manual.pdf
[2017/02/19]

ck technologies, "Supplier Manual, MT-003 8-27-13", 2013
http://www.cktech.biz/sites/cktech.biz/files/documents/ckt-supplier-manual-2013-08-27.pdf
[2017/02/18]

cooper industries, "Supplier Manual, C R P - S S - 0 1 - 0 1 - 0 1"
http://www.cooperindustries.com/content/dam/public/Corporate/Company/Sourcing/Supplier_Manual_English.pdf
[2017/02/18]

Crosby, P. B.,"Quality management: the real thing", Philip Crosby Associates II, Inc, 1(3), 1962

Crosby, P. B., "Development and Control of the Pershing Quality Program", SAE Technical Paper, No. 640559, 1964

Crosby, D. C., "Zero Defects vs. Six Sigma - The differences may surprise you", 2006
http://www.qualitydigest.com/inside/six-sigma-article/zero-defects-vs-six-sigma.html#

[2017/02/03]

CSQA, "Zero Defects - Do you agree?",
http://csqa.info/zero-defects
[2017/01/28]

Da Silva, M. G., Giasolli, R., Cunningham, S., DeRoo, D., "Mems design for manufacturability (dfm)", Sensors Expo, Boston (USA), 2002

Douglas Autotec Corp., "Supplier Manual, Q7.4.1.2-C-PU-WI-001 Rev. F 2-10-13", 2013
http://www.douglasautotech.com/wp-content/uploads/DAC_Supplier_Manual.pdf
[2017/02/18]

Dynapac Atlas Copco Group, "Dynapac Supplier Manual 2016", 2016
https://www.dynapac.com/assets/images/uploads/downloads/Dynapac_Supplier_Manual_2016.pdf
[2017/02/18]

Dynapar Corp., "Supplier Quality Manual Q-POL-451", 2015
https://www.dynapar.com/uploadedFiles/_Site_Root/About_Dynapar/Dynapar-Supplier-Quality-Manual.pdf
[2017/02/18]

EagleBurgmann, „EagleBurgmann Lieferantenbewertung"
http://www.eagleburgmann.de/media/admin-area/web/general-group/procurement/lieferantenbewertung_erlauterung_va.pdf/download
[2017/02/16]

Eaton Aerospace, "Eaton Aerospace Supplier Quality Manual, Rev. H"
http://www.eaton.com/ecm/groups/public/@pub/@eaton/@corp/documents/content/ct_256550.pdf
[2017/02/19]

ElringKlinger, " Zentrale Anweisung ZA QM 7.4-1", 2011
https://www.elringklinger.de/sites/default/files/dd_files_paragraph/1304_za_qm_7.4-1-supplier-evaluation.pdf
[2017/02/16]

Fisher & Company, "Supplier Manual 2015 Revision 2", 2015
http://www.fisherco.com/files/2015_Fisher_&_Company_Supplier_Manual_Rev2.pdf
[2017/02/18]

Fouch, G. E., "A Guide to Zero Defects. Quality and Reliability Assurance Handbook 4155.12-H.", Corporate Author : ASSISTANT SECRETARY OF DEFENSE
http://www.dtic.mil/dtic/tr/fulltext/u2/a950061.pdf
[2017/02/02]

Freudenberg NOK, „Freudenberg-NOK Supplier Manual WI-06-PURC-000", 2016
https://www.fst.com/-/media/files/suppliers-portal/nok/supplier-manual-rev-14-dated-january-2016.ashx
[2017/02/18]

Freudenberg Schwab Vibration, „Lieferantenbewertung; Anlage13 zur Qualitätsrichtlinie für Lieferanten D_07_33_01_TK", 2013
https://www.freudenberg-pm.com/-/media/Files/www,-d-,freudenbergpm,-d-,com/Suppliers/FPM_QM%20-%20Leitlinien%20fuer%20Lieferanten_DE.pdf
[2017/02/16]

Halpin, J. F., "Zero defects: a new dimension in quality assurance", McGraw-Hill, 1966

Handtmann, " Merkblatt „Lieferantenbewertung" Erläuterung der Bewertungskriterien", 2015
https://www.handtmann.de/fileadmin/user_upload/Geschaeftsbereiche/Leichtmetallguss/Einkauf/Downloads/handtmann-Merkblatt-001.pdf
[2017/02/16]

Honeywell, "Honeywell ACS Vendor Scorecard Version 2.1", 2008
https://sensing.honeywell.com/vendor-scorecard.pdf
[2017/02/16]

Ideal Automotive, "Beschreibung Lieferantenbewertung Produktionsmateriallieferanten B009", 2016
http://www.ideal-automotive.com/wp-content/uploads/2015/02/Beschreibung_Lieferantenbewertung_B009_de-1.pdf
[2017/02/16]

IER Fujikura, "Supplier Manual, IER-QSF 003", 2013
http://ierfujikura.com/wp-content/uploads/2016/01/IER-QSM-003-Supplier-Manual-Rev-4.pdf
[2017/02/18]

i2 tec Innovative Injection Technology, "Supplier Quality Manual, PU-003 Rev. 6", 2014
http://www.i2-tech.com/filesimages/pdfs/supplier-quality-manual.pdf
[2017/02/19]

Jochem, R., Raßfeld, C., "Qualitätsbezogene Kosten", in Masing Handbuch Qualitätsmanagement (Pfeifer, T., Schmitt, R.), 6. Auflage, Hanser, 2014

Junping, W., Dan, X., Yongbang, S., "A method for timing constrained redundant via insertion", Journal of Semiconductors, 35(4), 045010, 2014

Kahng, A. B., Samadi, K., "CMP fill synthesis: A survey of recent studies", IEEE Transactions on Computer-Aided Design of Integrated Circuits and Systems, 27(1), 3-19., 2008

Kalinowska, D., Kloas, J., Kuhfeld, H., Kunert, U., "Aktualisierung und Weiterentwicklung der Berechnungsmodelle für die Fahrleistungen von Kraftfahrzeugen und für das Aufkommen und für die Verkehrsleistung im Personenverkehr (MIV)", Endbericht, DIW Berlin, April 2005
http://www.diw.de/documents/dokumentenarchiv/17/44088/ModellaktEndbericht.pdf
[2017/02/04]

Kärcher, „Qualitätsmanagement – Leitfaden für Lieferanten", 2010
https://supplierinfo.kaercher.com/lieferantenleitfaden/deutsch/leitfaden.pdf
[2017/02/23]

Kim, S., Nakhai, B., "The dynamics of quality costs in continuous improvement", International Journal of Quality & Reliability Management, 25(8), 842-859, 2008

Kirchhoff Van Rob, „ Supplier Development Manual", Rev 29, April 14, 2016
http://www.van-rob.com/supplier/manual.pdf
[2017/03/02]

Klein, H.-P., Hediger, C., Harnisch, T., Subrahmanyan, R., "Flex PCB Reliability: An Objective Evidence based Approach", 2015
https://www.researchgate.net/profile/Torsten_Harnisch2/publication/272147992_Flex_PCB_Reliability_An_Objective_Evidence_based_Approach/links/555c539f08ae8f66f3adec5c.pdf
[2017/10/22]

Knott, „Handbuch für unsere Zulieferer", 2015
https://www.knott.de/wp-content/uploads/Knott_Lieferantenhandbuch_071015.pdf
[2017/02/16]

Kraftfahrtbundesamt, „Durchschnittliche Fahrleistung auf Vorjahresniveau", 2016
http://www.kba.de/DE/Statistik/Kraftverkehr/VerkehrKilometer/2015_vk_kurzbericht_pdf.pdf?__blob=publicationFile&v=9
[2017/02/04]

Kunert, U., Radke, S., Chlond, B., Kagerbauer, M., „Auto-Mobilität: Fahrleistungen steigen 2011 weiter", DIW Wochenbericht Nr. 47.2012
https://www.diw.de/documents/publikationen/73/diw_01.c.411737.de/12-47-1.pdf
[2017/02/04]

Leopold Kostal GmbH & Co. KG, „Kontakt Systeme Kostal", 2013
http://www.kostal.com/kks/download/KAE/6_Lieferantenbewertung_DE.pdf
[2017/02/16]

Li, X., Qin, J., Bernstein, J. B., „Compact modeling of MOSFET wearout mechanisms for circuit-reliability simulation", IEEE Transactions on Device and Materials Reliability, 8(1), 98-121., 2008

Liebmann, L. W., Mansfield, S. M., Wong, A. K., Lavin, M. A., Leipold, W. C., Dunham, T. G., "TCAD development for lithography resolution enhancement", IBM Journal of Research and Development, 45(5), 651-665, 2001

Liebmann, L. W., "Layout impact of resolution enhancement techniques: impediment or opportunity?", In Proceedings of the 2003 international symposium on Physical design (pp. 110-117). ACM, April 2003

Lin, C. W., Tsai, M. C., Lee, K. Y., Chen, T. C., Wang, T. C., Chang, Y. W., "Recent Research and Emerging Challenges in Physical Design for Manufacturability/Reliability", In Proceedings of the 2007 Asia and South Pacific Design Automation Conference (pp. 238-243). IEEE Computer Society, Jan 2007

Littelfuse, „CHI-10SDE-001-A Littelfuse Supplier Quality Manual Rev A", 2014
http://www.littelfuse.com/~/media/files/littelfuse/technical-resources/documents/supplier-quality/littelfusesqm-pdf.pdf
[2017/02/24]

Lu, H., Bailey, C., Yin, C., "Design for reliability of power electronics modules", Microelectronics reliability, 49(9), 1250-1255, 2009

Mack, C. A., "Fifty years of Moore's law", IEEE Transactions on semiconductor manufacturing, 24(2), 202-207, 2011

Magna Electronics, "Magna Electronics Supplier Quality Requirements Manual Rev 2"
http://www.magna.com/docs/default-source/suppliers/magna_supplier_quality_requirements_manual_rev2.pdf?sfvrsn=2
[2017/02/16]

MAN Nutzfahrzeuge Gruppe, "Lieferantenentwicklung MAN Nutzfahrzeuge Gruppe", 2006
http://www.man.eu/man/media/de/content_medien/doc/global_corporate_website_1/unternehmen_1/lieferantenentwicklung.pdf
[2017/02/23]

Mazotti, W., "Altera's Automotive Quality Program Building Higher Quality Automotive Electronics", 2016
https://www.altera.com/content/dam/altera-www/global/en_US/pdfs/literature/br/alteraautomotivequalityprogram.pdf
[2017/01/26]

McNulty, J. C., Knadle, K., Klein, H.-P., Harnisch, T., Dona, E., Meskens, W., Op de Beeck, M., Coder, "Assessment of Reliability Standards for Implantable Medical Devices", 2016
https://www.researchgate.net/profile/Torsten_Harnisch2/publication/282847265_Assessment_of_Reliability_Standards_for_Implantable_Medical_Devices/links/561e1cac08

aec7945a253f4d/Assessment-of-Reliability-Standards-for-Implantable-Medical-Devices.pdf
[2017/10/22]

Merrill Technologies, "Supplier Requirements Manual M M 0 3", 2015
http://merrillaviation.com/images/uploads/Merrill-Technologies-Group-Supplier-Requirements-Manual.pdf
[2017/02/19]

Mettas, A., "Design for reliability: Overview of the process and applicable techniques", International Journal of Performability Engineering, 6(6), 577-586, 2010

Milsco, "Supplier Quality Requirements Manual, Document 4-84-010 - Rev. D", 2014
http://www.milsco.com/Documents/SupplierQualityRequirementsManual.pdf
[2017/02/18]

MTU Friedrichshafen, „LBS Lieferanten Bewertungssystem", 2014
https://www.mtu-online.com/fileadmin/fm-dam/mtu-global/downloads/TP11_Handout_Lieferantenbewertung_MTUFN.pdf
[2017/02/16]

Nahar, A., Butler, K. M., Carulli, J. M., Weinberger, C., "Quality improvement and cost reduction using statistical outlier methods", In Computer Design, 2009. ICCD 2009. IEEE International Conference on (pp. 64-69). IEEE, Oct, 2009

Narasimhan, R., "Fulfilling quality requirements for automotive application", Global Semiconductor & Electronics Forum Connecting Technology Leaders, Hong Kong, March 8 – 10, 2017

NCR, "Supplier Quality Manual Document Number: 497-0469744 Revision: E", 2015
https://www.ncr.com/sites/default/files/ncr-supplier-quality-manual.pdf
[2017/02/18]

Norgren, "Supplier Performance Manual", 2015
http://cdn.norgren.com/pdf/N-LOG-00020.pdf
[2017/02/16]

Northrop Grumman Aerospace Systems, "Supplier Scorecard Guidelines SG-0110", 2016
http://www.northropgrumman.com/suppliers/OasisDocuments/Supplier_Scorecard_Guidelines.pdf
[2017/0216]

Northstar Aerospace, "Supplier Quality Manual MAN-QLT-01", 2016
http://www.nsaero.com/sites/default/files/MAN-QLT-01-supplier-manual-milton.pdf
[2017/02/18]

ON Semiconductor, "Effective Automotive Quality"
http://www.onsemi.com/pub_link/Collateral/TND387-D.PDF
[Nov 27th, 2015]

Pan, D. Z., Yu, B., Gao, J. R., "Design for manufacturing with emerging nanolithography. Computer-Aided Design of Integrated Circuits and Systems", IEEE Transactions on, 32(10), 1453-1472, 2013

Polte, T., Aal, A., "Quality per design. Dynamic demands enforces tailored answers", Volkswagen AG, Electronic Analysis and Robustness, Presentation at Infineon Quality Day, 2014

PTM Corp., "Supplier Manual", 2014
http://www.ptmcorporation.com/wp-content/uploads/PTM-Supplier-Quality-Manual.pdf
[2017/02/18]

PwC, "Spotlight on Automotive PwC Semiconductor Report", Technology Institute Interim Update Global Semiconductor Trends – Special Focus Automotive Industry. September 2013
http://www.pwc.at/images/tmt-studie-2.pdf
[2017/03/03]

Pyzdek, T., Keller, P., "The Handbook for Quality Management: A Complete Guide to Operational Excellence: A Complete Guide to Operational Excellence", McGraw Hill Professional, 2012

QCC, "Supplier Quality Assurance Manual, SP0607, Rev. E", 2011

http://www.qccorp.com/sites/default/files/QCC_Supplier-Quality-Assurance-Manual.pdf
[2017/02/18]

Rappitsch, G., Seebacher, E., Kocher, M., Stadlober, E., "SPICE modeling of process variation using location depth corner models", IEEE Transactions on Semiconductor Manufacturing, 17(2), 201-213, 2004

Sage Automotive Interiors, "SAGE Automotive Interiors Supplier Manual", 2013
http://www.sageautomotiveinteriors.com/pdf/SAGE%20_SUPPLIER_MANUAL_091913v2.pdf
[2017/02/16]

samtec, "samtech Supplier Quality Assurance Manual; CO-XX-ML-2003-M Rev. D", 2017
http://suddendocs.samtec.com/standard_products/quality_information/sqam_manual.pdf
[2017/02/19]

Saraswat, K., "Trends in Integrated Circuits Technology", Department of Electrical Engineering, Stanford University
https://web.stanford.edu/class/ee311/NOTES/TrendsSlides.pdf
[2017/03/16]

Schäffler, „Lieferantenbewertung / Supplier Evaluation nach / acc. QSV / QAA S 296001", 2006
http://www.schaeffler.se/remotemedien/media/_shared_media/12_suppliers/quality/quality_assurance_agreement__qaa_/supplier_evaluation_/5_1_Bewertungskriterien_Lieferantenbewertung_de_en_20061121.pdf
[2017/02/23]

Schiffauerova, A., Thomson, V, "A Review of Research on Cost of Quality Models and Best Practices", International Journal of Quality and Reliability Management, Vol, 23, No. 4, 2006

Schlemmer, „Supplier Rating in the Schlemmer Group"
https://www.schlemmer.com/files/leaflet_supplier_rating_in_the_schlemmer_group.pdf
[2017/02/18]

sl-america, "Supplier Requirements Manual, Document 10282013, Revision: - 01-15-14 *", 2014
http://www.sl-america.com/wp-content/uploads/SUPPLIER-REQUIREMENTS-MANUAL-01-15-14.pdf
[2017/02/18]

Steierwald, G., Künne, H.-D., „Stadtverkehrsplanung Grundlagen – Methoden – Ziele", Springer Verlag Berlin, Heidelberg, 1994

Strattec, "Supplier Quality Manual 06L2M002 Rev. AG (12/14)", 2014
http://www.strattec.com/files/6314/2204/6401/Supplier_Quality_Manual.pdf
[2017/02/18]

Suárez, J. G., "Three experts on quality management: Philip B. Crospy, W. Edwards Deming, Joseph M. Juran", TQLD Publication, (92-02), 1992

Sullivan, L., "Reducing Variability", in "The ignition switch from hell", Partnership 2000 LLC, 2015

TASUS Corp., "Supplier Quality Manual, Control # TS 7.4.1.B REV H"
http://www.tasus.com/wp-content/uploads/TASUS_Supplier_Manual.pdf
[2017/02/19]

TBDN Tennessee Company, "Supplier Quality Assurance Manual", 2014
http://www.tbdn.com/supplierquality/Forms/SQAM%2012.2.14.pdf
[2017/02/18]

Textron Systems, "AAI Corporation Textron Systems Supplier Rating System QA-SP48 Rev. E", 2015
http://www.textronsystems.com/sites/default/files/QA-SP48.pdf
[2017/02/19]

Thai Summit America, "Supplier Quality Manual, ROM-Q.7.4.1.2. Rev. 5", 2013
http://thaisummit.us/wp-content/uploads/2014/07/New-Supplier-Quality-Manual.pdf
[2017/02/19]

Thermotech, "Supplier Quality Manual, Revision J", 2016

http://www.thermotech.com/wp-content/uploads/2016/03/Thermotech-Supplier-Quality-Manual-1.pdf
[2017/02/19]

Thyssen-Krupp, " Supplier Rating Code CD-00030-EN", 2010
http://www.thyssenkrupppresta.com/Public/SupplierNet.nsf/e8d613ade919f37441256c61002bda31/1e356cda011b5c6341256e7b004ccc6a/$FILE/CD-00030-EN-v4.pdf
[2017/02/16]

TopWorx, Inc., "Supplier Performance Manual REVISION M", 2016
http://www2.emersonprocess.com/en-US/brands/topworx/supplierportal/Documents/Supplier%20Performance%20Manual.pdf
[2017/02/16]

Trehan, R., Sachdeva, A., Garg, R. K., „A comprehensive Review of Cost of Quality", International Journal of Research, Vol. 6, Issue 1, 2015

Tribar Manufacturing LLC., "Supplier Quality Requirements Manual, Rev. 01", 2014
http://www.tribarmfg.net/files/Supplier-Manual-Tribar.pdf
[2017/02/19]

T-online, "Kraftfahrtbundesamt veröffentlicht erstmals Statistik Ein Auto fährt pro Jahr im Schnitt rund 14.000 Kilometer"
http://www.t-online.de/auto/news/id_74583174/so-viele-kilometer-faehrt-ein-auto-im-durchschnitt.html
[2017/02/04]

United Technologies, "Supplier Quality Manual, Exhibit 1", 2014
https://files.ccs.utc.com/ccs/en/worldwide/contentimages/SQM-BIS-Final-2014-09-25.pdf
[2017/02/19]

Van-Rob Inc Kirchhoff Automotive, „Supplier Development Manual, REV: 029", 2016
http://www.van-rob.com/supplier/manual.pdf
[2017/02/18]

Vaxevanidis, N. M., & Petropoulos, G., "A literature survey of cost of quality models", Journal of engineering, 6(3), 274-283, 2008

Voith, " Lieferantenbewertung und Eskalationsprozess Juni 2012 VN 3207", 2012
http://voith.com/de/VN_3207_de.pdf
[2017/02/25]

Volvo Group, "Supplier Quality Assurance Manual, third edition", 2014
http://www.volvogroup.com/SiteCollectionDocuments/suppliers/attachments/PQP/SQAM%20V2014%20R5.0%20FINAL.pdf
[2017/02/19]

Volvo Group, "Supplier Quality Assurance Manual, fourth edition", 2016
http://www.volvogroup.com/content/dam/volvo/volvo-group/markets/global/en-en/suppliers/our-supplier-requirements/KEP-2_SQAM-2016.pdf
[2017/03/02]

Wang, K. S., "Towards zero-defect manufacturing (ZDM)—a data mining approach", Advances in Manufacturing, 1(1), 62-74, 2013

Wang, W., Reddy, V., Krishnan, A. T., Vattikonda, R., Krishnan, S., Cao, Y., "Compact modeling and simulation of circuit reliability for 65-nm CMOS technology", IEEE Transactions on Device and Materials Reliability, 7(4), 509-517, 2007

Wilbert Plastic Services, "Supplier Manual, Rev. 5", 2013
http://wilbertplastics.com/wp-content/uploads/2016/03/WPS-Supplier-Quality-Manual-Rev-5_1.pdf
[2017/02/19]

Wilkins, J. D., "The Bathtub Curve and Product Failure Behavior Part One - The Bathtub Curve, Infant Mortality and Burn-in"
http://www.weibull.com/hotwire/issue21/hottopics21.htm
[2017/05/01]

Wilkins, J. D., "The Bathtub Curve and Product Failure Behavior Part Two - Normal Life and Wear-Out", 2002
http://www.weibull.com/hotwire/issue22/hottopics22.htm
[2017/05/01]

Wittmann, J., Bergholz, W. "Introduction to Quality Management in the Semiconductor Industry: Volume I: General", CreateSpace Independent Publishing Platform, 2016

WKW, „Arbeitsanweisung SAP Lieferantenurteilung Zukaufteile & Lohnbearbeitung, AA 11-05-d-SAP", 2016
https://www.wkw.de/fileadmin/content/alt/bilder/teaserboxen/AA11-05-d-SAP_Lieferantenbewertung.pdf
[2017/02/16]

Yazaki Australian Arrow Pty. Ltd., "Supplier Rating Scheme", 2013
http://www.australianarrow.com.au/purchasingpdfs/004_Supplier_Rating_Scheme.pdf
[2017/02/16]

Zollondz, H.-D., "Grundlagen Qualitätsmanagement Einführung in Geschichte, Begriffe, Systeme und Konzepte", 2. Auflage, Oldenbourg Verlag München, 2011

ZVEI, „Handbook for Robustness Validation of Semiconductor Devices in Automotive Applications", ZVEI - Zentralverband Elektrotechnik- und Elektronikindustrie e. V. (German Electrical and Electronic Manufacturers' Association), Electronic Components and Systems Division, 2007

ZVEI, „Handbook for Robustness Validation of Automotive Electrical/Electronic Modules", ZVEI - Zentralverband Elektrotechnik- und Elektronikindustrie e. V. (German Electrical and Electronic Manufacturers' Association), Electronic Components and Systems Division, Revision: June 2013

www.ingramcontent.com/pod-product-compliance
Lightning Source LLC
Chambersburg PA
CBHW051159220526
45473CB00003B/825